美文工作室

站在 人 这边

思考步法
A-Z
AI时代元认知入门

Thinking Moves A-Z
Metacognition Made Simple

[英] 罗杰·萨特克利夫

[英] 汤姆·比格尔斯通

[英] 贾森·巴克利　　著

儿童哲学·中国　　译

上海教育出版社
SHANGHAI EDUCATIONAL
PUBLISHING HOUSE

序言

所有的教学都教授思考。即使是死记硬背的学习也会教授记忆技巧及个别事实。在这本书中，我们鼓励将思考"大胆说出来"——关注学校和日常生活中每天都在使用的思考模式，使学生成为大师级别的思考者（即能熟练地为自己手头上的任务挑选正确的思考工具）。总之，思考步法 A—Z[①] 是元认知发展的一个框架。

为了便于学习，我们按照字母表从 A—Z 的顺序编排了 26 个思考步法，在词汇的选择上，我们尽量遵循"金发姑娘原则"：所选用的步法名称是语义足够具体、可以直接使用的独立词汇，同时它们也具备了一定的灵活性和通用性。例如，我们

[①] 将 Thinking Moves 译为"思考步法"原因有二：第一，这里的 Moves 强调一种可重复的方式方法，具有一定的规则性和基础性，如同下棋的步法，"步"强调了行动，而"法"强调了规则；第二，Moves 还强调了一种动态性和组合的灵活性，如同舞步的各种步法，可以灵活搭配，碰撞出不同的火花。——译者注

使用了"**关联**"而不是含义更广泛的"关系";使用了"**推测**"而不是含义较为具体的"演绎"。

通过实践,我们确保了每一个思考步法都是一个能够直接进行的"思考动作",而非一种指向需要长期培养的性格特征或品质,比如"**提前思考**"或"回顾",而不是"适应力"或"同理心"。我们希望尽量避免读者在阅读后出现这样的回应:"看上去很不错,但在实际操作中会是什么样的呢?"思考步法是能够直接使用的,因此学生可以很容易地在自己的思考实践中找到相应的例子,或是能用简洁的语言来表述它们。

人类、社区乃至整个社会都因思考而富于变化,或进或退,或循环往复……思考是充满力量的。这本书帮助教师探讨如何推动年轻人的思考,让他们在应对挑战和机遇时变得积极且富有成效。

这本书的编排结构及其易于理解的特点,意味着思考步法将使教师能够将思考以建设性和连贯的方式融入他们不断发展的实践,从而使深入思考变得自然。通过这种方式,思维就会动起来……它会推动年轻人的意识发展、理解力和学习力提升。

——米克·沃特斯

教育学教授

今天的教育工作者们了解元认知、思维技能和学术词汇的重要性。为了让学习变得有意义，学生需要了解并使用思维词汇来综合自己的知识、价值观和技能。在《思考步法 A—Z：AI 时代元认知入门》一书中，结合克拉克斯顿、科斯塔和理查特的著作，萨特克利夫和他的团队为我们提供了一套必要工具。在这本清晰、使用者友好的指南中，每一个思考步法都经过了仔细检核和分析，辅以一组对应的同义词和一系列的游戏活动来进行拓展。每一位希望学生能够收获有意义的学习体验的教师，都可以借助这套优秀的工具包，为学生提供思维词汇支持，发展他们的元认知水平。

——特雷西·史密斯
学校校长

Preface

All teaching teaches thinking. Even rote learning teaches remembering as well as individual facts. What we encourage in this book is simply teaching the thinking "out loud" — drawing attention to the Thinking Moves already in use in school and in everyday life, so that students can become master thinkers, able to choose the right tool for the job in hand. In short, Thinking Moves A-Z is a framework for developing metacognition.

We have used the device of an A-Z of 26 Thinking Moves for ease of learning, and because that turns out to be a "Goldilocks" level of detail—specific enough to be directly useful, general enough to be flexible. So, for example, CONNECT is preferred to the broader "relationship"; INFER is preferred to the more specific "deduction".

We have made sure they are all workable "Moves", such as thinking AHEAD or BACK, rather than longer-term dispositions or character traits, such as resilience or empathy. We wanted to avoid the response, "That's a good idea. But what does it look like in practice?" Instead, we wanted the moves to be practical already, so that self-contained examples can be readily found in students' own thinking or expressed in a few sentences.

Thinking moves. It moves people, communities and societies ... forwards, backwards, sideways ... even round in circles. This book helps teachers to come to terms with how to move thinking in young people so that it is productive and positive in addressing challenge and opportunity.

The structure of the book and its accessibility mean that the approaches described will allow teachers to fit thinking constructively and coherently into their evolving practice with the result that it becomes natural to think deeply. In this way, thinking moves ... it moves awareness, understanding and learning in young people.

— **Mick Waters, Professor of Education**

Today's educators know the importance of metacognition, thinking skills and academic vocabulary. In order for learning to be meaningful, pupils need to acquire and utilise a thinking vocabulary with which to synthesise their knowledge, values and skills. Incorporating the work of Claxton, Costa and Richhart, in Thinking Moves A-Z: Metacognition for Everyday Learning, Sutcliffe and his team have provided us with an essential tool for this. Each Thinking Move is examined and analysed, alongside a set of synonyms and activities in a clear and user-friendly guide, so that any teacher who wishes their students to have meaningful learning experiences can provide them with the help of this excellent toolkit for providing thinking vocabulary and metacognition.

— **Tracey Smith, Headteacher, New Marston School**

引言

　　需要先向读者说明的是，这本小书主要面向的读者是那些认同元认知发展对学习的价值，认可并愿意尝试实践"思考步法 A—Z"框架的教师。

　　但是，本书同样也适用于其他任何想要了解元认知领域、掌握简便且实用的元认知工具的读者，尤其是专业和个人发展与元认知领域相关的朋友。

　　不妨让我们从对元认知的一个常见释义开始——"关于思考的思考"。

　　虽然从某种程度上来说，该释义有其可取之处，但它总体上带有误导性，因为该释义中第一个"思考"所指代的含义是模糊的，它处于以下两个含义之间的模糊地带：

- 某人在想什么（或已经想了什么）
- 某人是如何思考的（或是如何一直在思考的）

　　简单来说，这两个含义之间的区别，就是关于人的思想和思维过程的区别。

更准确地说，元认知其实是"关于思维过程的思考"，正如《牛津英语词典》中对于元认知的释义所指出的：

（元认知是）对某人的思维过程的觉察和理解，被认为尤其是对引导思维过程的方向有重要作用。

该定义的第一部分表明，产生想法的过程是千差万别的，我们首先需要觉知并识别出思考正在经历的不同过程，而第二部分则提示了元认知如何引导人们更好地管理自己的思维过程。

更好地管理自己的思维过程，从某种意义上来说是进行个人情绪管理和行动规划的关键，也即上文提到的个人发展。

如果回到教学语境，在正规教育中，我们把对思考的自我管理（尤其是个人情绪和行为管理）称为"自我调节"（有时也被称为"自我调节式的学习"——尽管这是一个相当狭义的概念）。

如果学生能够调节或控制自己的情绪（比如愤怒或不耐烦）和行为，既可以避免对他人的学习注意力产生影响，也是在为自己的自主学习做好准备。

这是一个理想化的有力想法。那么，学生应该如何成为一名善于自我调节的学习者，情绪和行为又将对思维过程的管理产生什么影响？

当然，实现这一目标知易行难。但我们至少可以先用即使小朋友都能理解的语言将之表达出来；这也是每个人的思维发展之旅中必不可少的一个阶段。

用思考步法的语言来表达，学生需要做的可能是：**聚焦-关注整体**（退后一步），**倾听**自己的感受（并意识到自己正处于愤怒或不耐烦的情绪之中），或是**审视**自己正在做的事情（及可能带来的影响，尤其是对他人的影响），并能够根据情况**变通**自己的思考和感受方式，调整自己的行为。

诚然，自我觉察及调节能力的培养需要时间，但明智的父母和教师正沿着这个方向给予孩子们更多的帮助。（值得一提的是，许多处在思维发展之旅中的成年人也需要获得这方面的帮助：帮助可以来自他人，比如通过某种咨询或治疗，也可以是自我助力，比如通过正念或冥想。）

比较有难度的是，我们能否用合适的步法语言来捕捉自己的思考、感受或行为，并在此基础上为自己选择更好的方式来进行思考、感受或行动。

但刚才提到的步法（**变焦、听/看**和**变通**）只是"思考步法A—Z"26个思考动作中的3个，在引导思考的过程中，其他的

23 个步法也都发挥着独特而重要的作用。

在某些情况下，步法使用的倾向性是非常明显的。例如，第一个步法——**提前思考**，很明显是关乎未来的，而第二个动作——**回顾**，则显然是指向过去的，因此，**提前思考**和**回顾**所适用的场景就非常清晰。（也有一些时候，人们需要提醒自己关注当下，这时候**听 / 看**的步法就很适用，因为它代表着所有感官的综合运用，它们只能专注于感受此时此地所发生的事情。）

在另一些情景中，当我们思考自己现在应该如何引导自己的思维过程时，答案就可能不那么明显。

你的思考过程可能是这样的：也许我应该试着把刚刚听到或想到的与之前听到或想过的**关联**起来，抑或是，我该试着看看它们之间的区别（**区分**）。或许我还能试着**解释**这种差异，或是直接提出一个自己的新想法（**构想**）。

所以这时，我们可能会选择继续进行所有这些步法……不过，如果想要获得更好的答案，现在最明智的选择就是深入阅读这本书！

还有一点需要强调：想实现元认知定义中"对某人的思维过程的觉察和理解"并能够实现引导及优化思维过程的效果，

我们必然要能先识别出这些思维过程，并能用一个合适的语言来阐述它们。

理想情况下，这个词汇表应该尽可能简单，这样即使是小朋友也能理解并据此进行练习；它还应该尽可能完整，这样就可以全面地涵盖人们的思维过程，随着年龄的增长，人们能够对自己的思维过程进行更全面的识别和更高质量的自我引导。

在罗列思维过程方面，前贤进行了不少尝试，包括布鲁姆于1956年创建的布鲁姆分类法，时至今日，它仍被用于区分和鼓励"更高层次"的思维。

但事实证明，尚无一个足够简单、能为高年级学生自主使用的词汇表，能作为元认知发展和自我管理的有效工具而被使用。

除了足够简单之外，可以说思考步法A—Z是目前最完整地涵盖了人类思维过程的词汇表。从字面上来看，思考步法按照从A至Z的顺序，将26个表示思考的动作的英文首字母与字母表对应起来，同时它还带有这样的隐喻：就像A—Z涵盖了完整的字母表一样，思考步法A—Z也涵盖了所有人类思维的基本过程。

仅凭此事实，就足以印证思考步法的口号"足够简单但同时又非常全面"。因此，即使读者只是先阅读了关于每个步法的简单阐述，也请记住一点，这些步法是已经存在的，是每个人在应对不同情况时常常调配使用的。A—Z 的排列顺序只是帮助我们更好地觉察并更有创造性地使用它们。

Introduction

To be clear from the start, this short book is intended primarily for teachers who have heard that metacognition is a "powerful lever of learning"[1], and that the A-Z scheme is a good, if not the best, way of putting it into practice.

But anyone else who wants a simple and practical introduction to the field of metacognition, especially as it relates to professional and personal development, will also find it helpful.

Let us begin with a common account of metacognition—"thinking about thinking".

This is good in part, but misleading overall. That is because the second reference to "thinking" is ambiguous as between:

---

1   Educational Endowment Foundation 2018 report, "Metacognition and Self-regulated Learning".

- WHAT one is thinking (or has thought)
- HOW one is thinking (or has been thinking)

In simpler terms, this is the distinction between one's thoughts and one's thought processes.

Metacognition, more precisely, is thinking about the latter, as the *Oxford English Dictionary* makes clear:

... awareness and understanding of one's thought processes, esp. regarded as having a role in the direction of those processes.

The first part of this definition suggests the need to *identify* different processes of thought, whilst the second hints at how this can lead one to *manage* those processes to better effect.

Better management of one's thinking is a key—perhaps the key—to better management of one's attitudes and actions: i.e. the self-development referred to above.

But, to return to the focus on teaching and learning, managing your thinking (and, importantly, your feeling and acting) is what in formal education is referred to as "self-regulation" (or sometimes "self-regulated learning"—though that is a rather narrower concept).

The idea is that if students learn to regulate or control their feelings (of anger, say, or impatience) and their behaviour, they will not only avoid distracting others from learning, but will be more ready to learn themselves.

It is a powerful idea and ideal. But how are students to achieve it, and what has it got to do with regulating or managing their *thinking*?

Achieving it is, of course, easier said than done. But what is involved can certainly be put into words that even young children can understand; and that is a essential stage of anyone's journey.

In Thinking Moves language, students need to ZOOM OUT (= step back) and either LISTEN to their feelings(that tell them they are angry or impatient) or LOOK at what they are doing (and its effect, especially on others)—and learn to VARY their way of thinking and feeling about the situation, and thereby their behaviour.

It does take time for this sort of self-awareness and self-correction to develop, but the best of parents and teachers are continually helping young people in this direction. (And, lest it be ignored, many an adult can still be helped—or help themselves—along the journey: some, via counselling or therapy of some sort, and some via mindfulness or meditation of some sort.)

In different Thinking Moves language, the challenge is to "catch yourself" thinking, feeling or acting, and to "coach yourself" to think, feel or act differently and better.

But the Moves just mentioned (ZOOM, LISTEN / LOOK and VARY) are just 3 of the 26 in the A-Z, and the other 23 all play different and important parts in enabling us to "direct" our thinking.

In some cases, the direction is very obvious. The first Move, for example—thinking AHEAD—points to the future, whilst the second Move—thinking BACK—points to the past. There are clearly times when one should do the former, and clearly times when one should do the latter. (There are also times when one should tell oneself to attend to the present. This is the Move, LISTEN / LOOK, which represents use of all of the senses—and they can only be used in the here and now.)

In other cases, the answer to the question, "how should I direct my thinking now?", might be less obvious.

Perhaps I should try to CONNECT what I have just heard or thought with what I had heard or thought before. Or perhaps I should try to see the difference (DIVIDE) between them. Perhaps I should try to EXPLAIN that difference, or perhaps I should just come up with (FORMULATE) a new idea of my own.

And so we might go on through all of the Moves… But your best way of proceeding now would be just to LOOK (read) further into the book!

There is just one final point of emphasis to make in this introduction: namely, that *"awareness and understanding of one's thought processes"*— and being able to direct them to good effect—surely does require one to *identify* them.

That, practically, means having a *vocabulary to label* them.

Ideally, this vocabulary should be as *simple* as possible, so that one can deliberately practice metacognition and self-regulation even as a child; and it should be as *complete* as possible, so that one can identify and implement the full range of thinking processes as one gets older.

There have been many attempts to list the fundamental processes of thought, including Bloom's famous taxonomy, created in 1956 and still used by some teachers to differentiate and encourage "higher order" thinking.

But no list has proved simple enough for even older students to *use effectively themselves*, i.e. as a tool for *being* metacognitive and self-managing.

And, we dare say, **no list is as complete as the A-Z**.

This scheme is literally (in English) an A-Z, with one Move per letter of the alphabet. But it is also *metaphorically* an A-Z. That is, it captures *all* the fundamental processes of human thought.

That fact alone justifies its slogan, *"Stunningly simple, but remarkably rich"*. So, even as you read the simple expositions of the Moves, try to keep in mind that they are deployed regularly by all of us in all sorts of contexts—often in clever combination with others. The A-Z just helps us deploy them more consciously and more creatively.

这本书是如何构成的

　　下一部分的表格展示了 26 个思考步法及与其对应的 2 个关键同义词。选择这些关键同义词，不仅基于它们能从概念上进一步阐明每个步法背后的意涵，也基于其在某些情况下可以展示步法意义上的略有不同的细微差别。当它们被记住并投入使用后（关于如何记忆思考步法，请参考本书后续章节），还可以代替原步法成为日常生活中使用的思考动作，进一步丰富我们的思维词汇。第二个表格为每个思考步法匹配了一个重要的指导问题，这些问题适用于任何情境。就像咨询师使用的问题一样，它们的目的不是提供建议，而是引导学生发掘出自己最佳的思考。

　　在本书的核心部分，每一个思考步法都被分为四个部分。第一部分包含了这个步法的图标、关键同义词、一些启发问题，以及易于学生理解的解释语言。剩下的部分则以教师的需求为主。

　　第二部分是对该思考步法的释义，提供了一系列与该思考步法相关的词汇表，其中前四个词是它的动词同义词，用于显示这个思考动作的范围。中间的主体列表是与该思考步法相关的其他词汇，既包括日常用语，也包括更具学术性的语言，它们都是优秀思考者词汇的一部分。列表中的最后几个词指的是刻意练习这一思考步法后所表现出的智力美德，例如经常"回顾"的人善于回忆和反思。这也是这套工具的一个重要价值指向，即希望帮助人们将思考步法的培养当作一种习惯的塑造或"心态"的练习。

接下来的页面上，我们提供了探索该思考步法的不同路径。为每一种步法配了与其相关联的插画。部分活动，尤其是每一步法的第一项，被设计成了可以单独实施的内容。在活动中，对思考步法的讨论，往往通过有趣的元素而非沉重的课程内容来开启。其他活动则是被设计成某一节课中的一部分内容，这样教师可以在课程中有计划地使用或随机使用。应用部分指出了思考步法在各学科中运用的可能性。学科中的名人语录则提供了灵感或思想碰撞。

在对 26 个思考步法进行逐个解释之后，是关于"思考径"的章节，"思考径"指由多个步法组成的思考步骤，实际上是可以教授给学生的策略。但学生也可以依据文本进行自主应用——在这个维度上，我们可以将之称为"元认知策略"。

接下来的部分是关于思考步法在学习和生活中的使用，以及其是如何与其他现有的工具体系联系起来的，如阿特·科斯塔的"思维习惯"。

我们希望思考步法可以在教学活动中协助教师赋能学生。除此之外，思考步法和思考惯例在职业生涯和日常生活中可能也在规划、创造和决策上有所助益。请与我们分享您在实践思考步法和思考惯例的过程中的新发现，以及学生对它们的反应。

这本书是如何构成的

How This Book Is Organized

A table in the next section presents the Moves, along with two key synonyms. These synonyms have been chosen to help clarify the concept behind each Move, but also, in some cases, to point to slightly different nuances of meaning. They can be used routinely instead of the Moves, once the scheme has been memorized and operationalized,as indeed can any other synonyms. In a second table, each Move is linked with a key coaching question, for use in any context. Like the questions used by counsellors, their object is not to give advice, but to draw out from students their own best thinking.

In the heart of the book, each Move is presented into four sections. The first section contains an icon for the Move, its key synonyms, a couple more elicitation questions, and an explanation in the sort of language you might use when introducing it to students. The remainder continues with the teacher in mind.

The Second section contains a list of connected vocabulary and the explanation of the Move. In the connected vocabulary, the first four words are other verb synonyms showing the range of the move. The main list is of other associated words that would be part of a good thinker's vocabulary, from the everyday to more academic language. The final words in that list point to the intellectual virtue shown by a person who deliberately practices the move. Someone who regularly thinks back is good at recollection and reflection. This is an important aspect of the scheme, designed to help people cultivate Thinking Moves as habits or "mindsets".

In the next section for each Move, we offer ways to explore it in practice. We add connected pictures for each Move. Some of these activities, in particular the first for each move, can be done in isolation. They often have an element of fun, to open up discussion of a move without the weight of curriculum content. Others deploy a move in the context of a lesson and can be planned or used opportunistically. Applications point to some of the possibilities across the curriculum; quotations provide inspiration or provocation.

Following the explanations of the individual Moves, there is a section on Thinking Grooves, i.e. sequences of Moves. Grooves are, in effect, strategies that can be taught to students, but which they can appropriate and use autonomously (in which case they may be referred to as "metacognitive strategies".)

There is also a section on using the Moves, and another one in which we summarize how the Moves connect to other existing schemes, such as Art Costa's Habits of Mind.

We hope you find these useful in your teaching and that they empower the students who learn from you. You may also find the Moves and Grooves an aid to planning, creativity and decision-making in professional and wider life. Please do share with us the new applications for the Moves and Grooves that you develop, and your students' reactions to them.

THINKING MOVES A-Z LISTS
思考步法 A—Z 及关键同义词列表

	THINKING MOVES 思考步法	**EVERYDAY SYNONYMS** 同义词 I	**ALTERNATIVE SYNONYMS** 同义词 II
A	AHEAD 提前思考	PREDICT 预测	AIM 确定目标
B	BACK 回顾	REMEMBER 记住	REFLECT 反思
C	CONNECT 关联	LINK 链接	LIKEN 比拟
D	DIVIDE 区分	SEPARATE 分开	LIST 罗列
E	EXPLAIN 解释	SAY HOW 说明如何……	CLARIFY 澄清
F	FORMULATE 构想	SUGGEST 建议	PROPOSE 提议
G	GROUP 归类	SORT 分理	CLASS 分门别类
H	HEADLINE 标题	SUMMARIZE 概括	DISTIL 提炼
I	INFER 推测	DEDUCE 演绎	TAKE FROM 提取
J	JUSTIFY 证明	GIVE REASONS 给出理由	ARGUE 论证
K	KEYWORD 关键词	HIGHLIGHT 突出重点	PINPOINT 准确定位
L	LISTEN/LOOK 听 / 看	NOTICE 留意	GATHER 收集信息
M	MAINTAIN 保持	BELIEVE 相信	AFFIRM 断言
N	NEGATE 否定	DISAGREE 不同意	OPPOSE 反对
O	ORDER 排序	SEQUENCE 列序	ARRANGE 安排
P	PICTURE 描绘	IMAGINE 想象	PUT YOURSELF 设身处地
Q	QUESTION 提问	ASK 询问	WONDER 好奇
R	RESPOND 回应	ANSWER 回答	REPLY 答复
S	SIZE 度量	ESTIMATE 估计	QUANTIFY 量化
T	TEST 试验（验证）	DOUBT 怀疑	CHECK 检核
U	USE 使用（工具）	TRY OUT 试用	APPLY 应用
V	VARY 变通	CHANGE 改变	ALTER 更改
W	WEIGH UP 权衡（选择）	DECIDE 决定	JUDGE 判断
X	eXEMPLIFY 举例	GIVE EXAMPLE 举例说明	ILLUSTRATE 图示
Y	YIELD 让步	ACCEPT 接受	CONCEDE 退让
Z	ZOOM 变焦（聚焦细节 / 关注整体）	FOCUS ON 聚焦	SURVEY 审视

THINKING MOVES COACHING QUESTIONS
思考步法与指导问题

THINKING MOVES 思考步法	COACHING QUESTIONS 指导问题
A AHEAD 提前思考	What do you think will happen? 你认为会发生什么？
B BACK 回顾	What were the last two ideas? 最后两个想法是什么？
C CONNECT 关联	How do those connect? 它们是如何连接的？
D DIVIDE 区分	How is that different? 这有什么不同呢？
E EXPLAIN 解释	How do you mean? 你是什么意思？
F FORMULATE 构想	What ideas have people got? 人们有什么想法？
G GROUP 归类	How would you sort these into groups? 你怎么把它们分成几组？
H HEADLINE 标题	How would you say that in one sentence? 你用一句话怎么说？
I INFER 推测	If that's true, what else is true? 如果这是真的，那么还有什么是真的？
J JUSTIFY 证明	Can you say why? 你能说说为什么吗？
K KEYWORD 关键词	Which five words are most important here? 哪五个词在这里最重要？
L LISTEN/LOOK 听 / 看	What do you notice? 你注意到了什么？
M MAINTAIN 保持	Who is a "yes"? 谁是"是"？
N NEGATE 否定	Who is a "no"? 谁是"否"？
O ORDER 排序	What's the best way to organize this? 最好的排列方式是什么？
P PICTURE 描绘	What do you see when you picture this? 当你在脑海中描绘这个的时候，你看到了什么？
Q QUESTION 提问	What's the juiciest question here? 这里最能启发思考的问题是什么？
R RESPOND 回应	What do you say to that? 你想对此说什么？
S SIZE 度量	What sort of number are we talking? 我们能试着估计一下有多少吗？
T TEST 试验（验证）	How could you tell if that's true? 你怎么知道那是真的？
U USE 使用（工具）	How can you use it? 你可以如何使用它？
V VARY 变通	How else could we think? 我们还能怎么想？
W WEIGH UP 权衡（选择）	Which choice has more back-up? 哪个选择有更多的支持？
X eXEMPLIFY 举例	Can you give me an example? 你能给我举个例子吗？
Y YIELD 让步	Can you disagree with yourself? 你可以反对自己的想法吗？
Z ZOOM 变焦（聚焦细节 / 关注整体）	What's the big/little picture? 大蓝图 / 缩略图是什么？

目 录
CONTENTS

Q
QUESTION
提问

/ 168

R
RESPOND
回应

/ 178

S
SIZE
度量

/ 188

T
TEST
试验（验证）

/ 198

U
USE
使用（工具）

/ 208

V
VARY
变通

/ 218

W
WEIGH UP
权衡（选择）

/ 230

X
eXEMPLIFY
举例

/ 240

Y
YIELD
让步

/ 250

Z
ZOOM
变焦（聚焦细节 / 关注整体）

/ 260

AHEAD

提前思考

松　鼠

Relating Animal: Squirrel

关联理由：为未来提前准备（过冬）

Relating Reason: Preparing for the future (winter)

01

WHAT IS *AHEAD* ?

Synonyms	Coaching Questions
Predict	What do you think will happen?
Aim	What are you aiming for?

You do this sort of thinking everyday—when you wake up, you're already thinking about what will happen that day. There are lots of ways to think ahead—you might be predicting what will happen next in a story, getting ready for something, setting yourself a challenge, or be looking forward expectantly to a special occasion.

01

什么是"提前思考"?

关键同义词	指导问题
预测	你认为会发生什么?
确定目标	你的目标是什么?

其实你每天都在进行"提前思考"——早上醒来的时候,你就已经在想这一整天将会发生什么了。"提前思考"有很多种方式:可能是预测故事接下来将发生的情节,可能是为某件事进行准备,也可能是去给自己设定一个挑战,或者期待一个特殊的时刻。

02

EXPLAIN *AHEAD*

Predicting, preparing, intending and hoping are all ways of bringing the future into the present. Anticipating what will or might happen enables us to be ready when it does. Foretelling dangers enables us to minimize harm, whilst foreseeing opportunities enables us to maximize benefits.

In school, thinking ahead is crucial for forming aims and targets, setting short-term goals for group work and preparing revision timetables. ("How should I organize these notes so they will help me revise?")

With young students, we do most of the thinking ahead for them, but getting them to think ahead for themselves is important for developing independence.
("What's needed in today's school bag?")

Synonyms	Alternative Synonyms		Intellectual Virtue
Look forward	Future	Precaution	Anticipation
Expect	Goal	Foresight	Resolution
Hope	End	Probable	
Target	Means	Possible	
	Ambition	Inevitable	
	Risk	Consequence	

02

"提前思考"步法释义

预测、准备、计划和希望都是将未来带入现在的方法。预测将要发生或可能发生的事情，使我们能够在它真正到来时做好准备。预见危险使我们能把危险带来的伤害降到最低，而预见机会使我们能将可能获得的利益最大化。

"提前思考"对于学校生活中常见的制定目标、设定小组活动短期目标和制定复习时间表等环节而言至关重要。

（"我该如何整理这些笔记，好让它们帮助我复习？"）

对于年幼的学生来说，父母和教师为他们做了大部分的"提前思考"的工作，但是让孩子自己进行关于计划和预测的思考，对于培养他们的独立性来说是很重要的。

（"今天的书包里需要准备些什么？"）

同义动词	相关词汇		智力美德
期待	未来	预防	预测
预计	目标	远见	解决办法
希望	结束	很可能发生的	
瞄准（目标）	意义	有可能发生的	
	志向	不可避免的	
	风险	结果	

03

ACTING PLAN

What Would Happen Then?

This storytelling game thinks ahead through chains of predictions. Start with an imaginative scenario, e.g. "What would happen if you were invisible?" Take the first suggestion offered, e.g. "You'd be lonely because nobody could see you." Then ask, "What would happen then, because of that?" Continue the chain, recapping occasionally or breaking into pairs to generate new ideas. "Because of that" is the key element, as events in stories are not random; they follow in a chain of consequences.

Won't, Could, Will

Get some predictions for what won't happen next (this helps to narrow the domain of reasonable answers) and what could happen next. Then, from the candidate "coulds" (four or five are enough), see if the students can agree, with reasons, an Order of likelihood. (Can they be certain of any event?)

Project Projections

Projecting one's thinking into the future is essential for a successful project. Next time there's a group task that lasts two lessons or more,

03

然后会发生什么?

这是一个讲故事的游戏,我们需要通过进行一连串的预测来完成提前思考。游戏将从一个想象的场景开始,例如:"如果你是隐形的,会发生什么?"想象这个场景,对问题进行回答,例如:"没有人能看到你,因此你会感到孤独。"然后继续进行提问和预测:"如果你感到孤独,会发生什么?"将游戏以此形式继续下去,教师可以进行总结,或者通过两两讨论的方式来帮助学生产出新的想法。"因此"是游戏的关键元素,因为故事中的事件不是随机的,它们会产生一连串的后果。

不会发生,可能会发生,将要发生

对接下来"不会发生"(预测"不会发生"的事情将有助于缩小合理答案的范围)和"可能发生"的事情进行预测。然后,从"可能发生"的事情中选择4—5个选项,看看学生是否能就这些事情发生的可能性进行**排序**并给出理由,通过讨论并最终

pause the students after the first lesson and ask them, "If you carry on like this, will you achieve what you want to achieve?" They can reflect on current progress and predict how they will fare against their deadline. They might need to change tack!

Applications

Applications
What won't / could / will happen ...?

Music
... later in a composition

ICT
... in a robot's response to an algorithm ...

Wise Sayings

If you can look into the seeds of time, and say which grain will grow and which will not, speak then unto me.

——Banquo in William
Shakespeare's *Macbeth*

Begin with an end in mind.

——Stephen Covey

达成共识。（在这些"可能发生"的事情中，能肯定有必然会发生的事件吗?）

项目预测

　　提前预想未来进展是完成一个成功项目不可或缺的环节。当进行某个需要持续两节课及以上时间的小组任务时，可以在第一节课后请学生暂停并询问："如果继续保持目前的做法，你们能达成想要的目标吗?"学生可以反思目前的进展，并预测自己最后的表现。他们也可能会在这个环节后意识到，接下来自己需要改变策略!

学科中的应用

应用

什么不会发生 / 什么可能会发生 / 什么将要发生?

音乐

在后来的作品中……不会发生 / 可能会发生 / 将要发生?

信息通信技术

在机器人对算法的回应中……不会发生 / 可能会发生 / 将要发生?

名人语录

如果你能看透时间的种子，请告诉我哪些谷粒会生长，哪些不会。

　　　　——威廉·莎士比亚《麦克白》

开始时心中就有了结局。

　　　　　　——史蒂文·科维

B

BACK

回顾

关联动物

大　象

Relating Animal: Elephant

关联理由：有良好的记忆力
Relating Reason: Good memory

01

WHAT IS *BACK*?

Synonyms	Coaching Questions
Remember	What can you recall from ...?
Reflect	Let's take some time to review/reflect!

You do a lot of this—especially after something enjoyable or exciting, like remembering a day out or a successful game. Of course, sometimes you may remember things that didn't go so well. This isn't necessarily a bad thing though—because it could help you to make things better in the future.

01

什么是"回顾"？

关键同义词	指导问题
记住	你能回忆起什么吗？
反思	让我们花点时间回顾／反思一下！

　　你会做很多类似"回顾"的行为——尤其是在一些令人愉快或兴奋的事情之后，比如结束外出的一天或一场成功的比赛后。当然，有时你可能会记得那些不太好的体验，这并不一定是一件坏事，因为它可以帮助你在未来做得更好。

02

EXPLAIN *BACK*

At its simplest, learning is organizing new information and ideas into our existing way of thinking. A lot of this process takes place unconsciously, but deliberate, regular reflection is the high road to good learning. If one move rules all, it is this, metacognitive, one.

It demands that you stop thinking (ironically!) and think again about the moves you have made in the past and could make in the future, not least the final move in the A-Z: **ZOOM**. You might decide to zoom out to see the big, general picture, or to zoom in on something in particular.

There are other ways of organizing and memorizing ideas. Converting information into stories or sequences (such as acrostics or an A-Z) can help, as can creating **PICTURES** in your mind. Psychologists recommend thinking back often on what you are trying to learn, as well as trying to **HEADLINE** or summarize it. They also say that sleeping well helps.

Thinking back can also be a way of dealing with present or future

02

"回顾"步法释义

　　简单地说，学习就是将新的信息和想法组织到我们现有的思维方式中。这个过程很多时候是在无意识中发生的，但深思熟虑、有规律的反思是通往高质量学习的捷径。如果让一个思考步法来"统治"其他所有步法，那它必然是最富有元认知色彩的"回顾"。

　　"回顾"要求你停下此刻的思考（多么令人意外！），反思自己在过去用到的步法，并进一步考虑未来可以使用的步法，尤其是结合 26 个思考步法中的最后一个："**变焦（关注整体 / 聚焦细节）**"之时。在回顾之后，你可能会决定到底究竟如何"变焦"，是从宏观角度来思考全局，还是从微观角度聚焦某些细节来思考。

　　有许多方式可以帮助我们组织并记忆想法。例如，将信息转换成故事或序列（藏头诗或按某种顺序，如 A—Z 的字母排列顺序），就像在你脑海中创造和**描绘**图像一样。心理学家建议，要经常回想你正在学习的内容，同时试着进行"标题"式概括

challenges. This is particularly important for nurturing resilience in the face of difficulty: once a challenge has been overcome, recalling how intimidating it seemed at the start can reduce the trepidation next time.

Synonyms	Alternative Synonyms		Intellectual Virtue
Recall	Beginning	Memory	Recollection
Rehearse	Origin	Recollection	Reflection
Think again	Past	Reminiscent	
Chew over	History	Replay	
	Ancestry	Second thoughts	
	Forerunner	Turning point	

或总结。此外，好的睡眠对此也非常有帮助。

回顾能帮助我们更好地处理当下及未来可能遇到的问题和挑战。回顾对培养抗压力尤其重要：克服挑战之后，回想一开始那令人生畏的情形，可以减少我们下一次应对新挑战时的恐惧。

同义动词	相关词汇		智力美德
回忆	开始	记忆	回忆
排练	起源	回忆	反思
再想想	过去	回忆往事	
仔细考虑	历史	重演	
	祖先	再想想	
	先驱	转折点	

03

ACTING PLAN

Remains of the Day

This is a memory game, but it also shows how accounts of the past are always selective and usually subjective. In a quick, individual exercise, each student writes down 10 things they remember happening during the previous school day that don't happen every week (so as to avoid just a list of times of the day). You can then compare the lists and notice the things that some remember but others do not.

Remains of the Lesson

Half way through a lesson or at the end, invite students to recall steps in the lesson so far. The steps can be simple events: *"How did the lesson start?" "What did I ask you to do then?" "What happened next?"* For more challenge, you can ask what people said, or just for "ideas". You or the students might represent the steps in a flow chart, as a record of the lesson.

03

步法应用

昨日记忆

这是一个记忆游戏，这个游戏也充分表明，人对过去的描述总是带有选择性与主观性。请学生独立思考昨天在学校中发生的事，然后快速写下 10 件昨日发生但并非每周发生的事（这样做是为了避免他们仅仅是列出一天的流水账）。请大家分享自己记录的事件列表，比较后我们会注意到一些人记得而另一些人没有记得的事情。

课程印象

在课程中间或结束时，请学生回忆课堂活动经历的步骤。教师可以从对步骤的简单提问入手："活动是怎样开始的？""接下来，我的问题是什么？""接下来发生了什么？"更有挑战性的做法是，询问学生在课堂活动中，其他人说了什么，或者仅是他们收获的"想法"。教师或学生可以将这些课堂步骤用流程表的形式记录下来。

Pause for Reflection/Reflective Journals

The importance of pausing occasionally during a lesson for a minute's private reflection is underestimated. Encourage students to have a reflective journal, in which they record not only the main ideas just introduced or discussed (see KEYWORD later), but also random reflections of their own about their experiences or understandings.

Applications

Applications
What happened ...

Drama
What happened in the previous scene?

PE
What happened the last time your team had a freekick?

History
What happened when we last learnt about Edward I?

Wise Sayings

Life must be lived forwards, but can only be understood backwards.

————Søren Kierkegaard

Without reflection, we go blindly on our way, creating more unintended consequences.

————Margaret J. Wheatley

停一停！反思／反思日记

人们低估了在课堂中偶尔停下来的价值。停一停！让我们来进行一分钟的个人反思。可以鼓励学生写一本"反思日记"，日记中不仅可以记录刚刚介绍或讨论的主要想法（见后文中出现的"关键词"），也可以记录他们在活动中对自己的经验或理解的随机反思。

学科中的应用

应用

……发生了什么？

戏剧

上一幕中发生了什么？

体育

你的球队上次获得任意球机会时发生了什么？

历史

我们上次学到爱德华一世时发生了什么？

名人语录

虽然生活要向前看，但只有常常回省，才能理解其中的奥义。

——苏林·克尔恺郭尔

没有反思而一直盲目地往前行，会带来更多意料之外的后果。

——玛格丽特·惠特利

C

CONNECT

关联

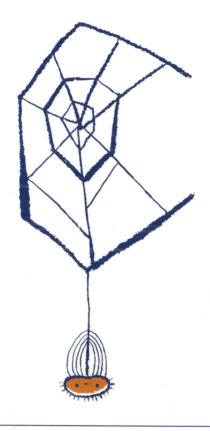

蜘　蛛

Relating Animal: Spider

关联理由：将事物（用网）关联在一起

Relating Reason: Connecting things (web) together

01

WHAT IS *CONNECT*?

Synonyms	Coaching Questions
Link	How are X and Y related?
Liken	How are X and Y similar?

Your brain has been making connections pretty much since you were born—linking one thing with another. Maybe you find them in the same place, or they happen at the same time, or they might be alike in some way. You can also link words or ideas together. Creative thinking often involves linking ideas that no one has put together before, such as "Paddington" and "bear".

01

什么是"关联"？

关键同义词	指导问题
链接	X 和 Y 有什么关系？
比拟	X 和 Y 有什么相似之处？

　　从出生起，你的大脑就一直在事物间建立关联——把一件事和另一件事链接起来。也许是因为你会在同一个地方发现它们，也许是因为它们同时发生了，也许是因为它们在某些方面可能是相似的。你也可能将词或想法链接在一起。创新式思维通常就包含将前人觉得没有关联的想法联系起来，比如"帕丁顿"和"熊"。

02

EXPLAIN *CONNECT*

The human brain seems primed to make an almost infinite number of connections, connecting one idea with another by spotting some relationship or similarity between them—even when presented with random pairs of words such as "melon" and "hedgehog".

Connecting ideas systematically is what makes for deep learning and understanding, rather than surface knowledge and rote definitions. For example, making connections helps learners understand the whole water cycle, rather than just the isolated definitions of the stages, because it is the inter-relationships between stages that are significant.

Extensive research suggests that such understanding can't happen from chalk and talk: learners must actively make the connections.

Synonyms	Alternative Synonyms		Intellectual Virtue
Put together	Similar	Comparison	Association
Associate	Alike	Comparatively	Assimilation
Match	Resemblance	Correlation	
Compare	Identical	Relevant	
	Relationship	Analogy	
	Common	Metaphor	

02

"关联"步法释义

人类的大脑似乎做好了建立近乎无限关联的准备。通过找出它们之间的关系或相似性来将一个想法与另一个想法联系起来——即使是像"甜瓜"和"刺猬"这样一对随机出现的词语。

深度学习和理解需要我们将想法系统地联系起来，而不是仅掌握表面知识和死记硬背定义。例如，学习"水循环"时，建立联系有助于学习者理解整个过程，单独地记忆、理解每个阶段的定义并不能起到真正的帮助，因为重要的是其中各阶段之间的相互关系。

广泛研究表明，深度理解无法通过单纯的讲授来实现：学习者必须积极主动地进行事物间的关联。

同义动词	相关词汇		智力美德
放在一起	类似的	比较	联想
联系	像	相对地	同化
匹配	相似之处	相关	
比较	完全相同的	有关的	
	有关系的	类比	
	共同的	比喻	

03

ACTING PLAN

Alike Like This

Get an idea for an object from a volunteer, e.g. a mug. Someone thinks of something that is like a mug in some way, such as, "A mug and a glass are alike like this—you can drink from both of them." The chain continues with a suggestion for something that is like a glass in a different way. For example, "A glass and window are alike like this—they're both breakable", then "A window and a hole in a wall are alike like this—you can see through them", "A hole in a wall and an injury are alike like this—they can both be caused by a bomb", and so on.

Missing Connection

Make two lists in columns of random nouns (concrete or abstract), perhaps suggested by students. List A should have 5-10 items, and List B 6-11, i.e. one extra. In pairs, students should copy the lists and connect (literally with a line) each item in A with one item in B, making sure they can explain the connections. Initially, pairs can share their results with each other informally. When all pairs are ready, see which item in B is hardest to connect with and perhaps why.

03

步法应用

有相似处

请一个学生提出一件物品，比如一个马克杯。让学生思考后提出一件其他物品，它需要在某个维度上与"马克杯"有相似之处。比如，马克杯和玻璃杯有相似处：它们都可以用来喝水。然后从"玻璃杯"开始继续建立链接：什么东西和"玻璃杯"有相似处？例如："窗户和玻璃杯有相似处：它们都是易碎的东西。""墙壁上的洞和窗户有相似处：都可以通过它们看到外面。""伤口和墙壁上的洞有相似处：它们有可能都是由爆炸引起的。"以此类推，将这个游戏继续下去。

迷失关联

在纸上写下两组随机的名词词单（可以是具体名词，也可以是抽象名词），名词可由学生自己选择。A词单包含5—10个名词，B词单包含6—11个名词，需要确保B组的名词数量比A组多一个。学生两人一组，抄写两组词单，尝试在B词单中为每个A组的名词找到一个与它能够产生关联的词，用直线连

Joined-up Thinking (Last Three to Speak)

This is a bit like *Remains of the Lesson* (see BACK) but with a focus on connecting ideas. After any sequence of three (significant) contributions by students, call for a pause and ask, "*Who were the last three to speak?*" Give students a little time to recollect, then put them in pairs to agree answers to "*what did each person say?*" and "*How did what they say connect?*" Repeated practice of this exercise will pay dividends in better dialog—more attentive listening and more constructive speaking.

Applications

Applications
How does ...

Art
How does Impressionism relate to Expressionism?

Philosophy
How does that question link to this one?

Math
How does this sum help us solve that equation?

Wise Sayings

Creativity is just "connecting things".

——Steve Jobs

Shall I compare thee to a summer's day?

——William Shakespeare

起来，并说出关联理由。学生通过两两对话，自由地分享想法。所有小组都完成后，一起看看 B 词单中的哪个词语最难关联，并尝试说出理由。

整体思维（刚才的三位发言人）

这个游戏和在"回顾"一节中出现的"课程印象"游戏有点相似，但它的重点在于联结想法。在任意三个学生连续发言（需是有意义的发言）之后，暂停并提问："最后三位发言的是谁？"给学生一点时间回忆，然后两人一组，回忆"每个人都说了些什么？"以及"他们的发言是如何关联到一起的？"反复练习这个游戏将会带来更高质量的对话——培养更专注的倾听，获得更有建设性的发言。

学科中的应用

应用
……如何

艺术
印象派和表现主义是如何关联的？

哲学
这两个问题之间是如何关联的？

数学
求和如何帮助我们解方程？

名人语录

创造力就是"将事物联系在一起"。
——史蒂夫·乔布斯

我可以把你比作夏日吗？
——威廉·莎士比亚

D

DIVIDE

区分

関联动物

剑 鱼

Relating Animal: Swordfish

关联理由：（将水面）一分为二

Relating Reason: Splitting things (water)

01

WHAT IS *DIVIDE*?

Synonyms	Coaching Questions
Separate	How are X and Y different?
List	Let's list the differences!

Before your brain can even make connections between things, it has to recognize that some things are different from others. It has to separate them out or differentiate them. Otherwise you would not be able to spot danger, for example, or find your food. And you would not hear different words, let alone understand them.

01

什么是"区分"？

关键同义词	指导问题
分开	X 和 Y 有什么不同？
罗列	让我们列出不同之处吧！

　　你必须先认识到事物的不同之处，才能开始在事物间建立联系。能够区分或者进行差异化是必要的步骤。否则，你就有可能无法识别危险，或者无法找到食物。你甚至无法听到不同的词语，更不用说理解它们了。

02

EXPLAIN *DIVIDE*

Telling the difference is one of the first things learners are asked to practice. Learning language itself involves a huge amount of differentiation of sounds, words and meanings. Speakers of one language can find it hard to notice that two sounds from another language are different. Students need to be trained in spotting all sorts of differences—between experiences, objects, situations, words, ideas—so that they can respond appropriately.

Careful analysis becomes more important as a learner progresses through the education system. In Science you learn to dissect things into parts or elements. In History little details matter, and in Art you become aware of finer features and aspects of sensory experience.

Synonyms	Alternative Synonyms		Intellectual Virtue
Tell apart	Different	Part	Differentiation
Distinguish	Opposite	Element	Dissection
Take apart	Distinction	Feature	
Analyze	Exception	Complex	
	Contrast	Binary	
	Whereas	Borderline	

02

"区分" 步法释义

　　区分差异通常是学习者首先要进行的事情之一。学习语言本身就涉及大量对音、词和意义的区分。单一语言的学习者会注意到自己很难发现另一语言中两个发音的差异。学生需要通过训练来识别各种差异——包括体验、物体、情境、词语、思想之间的差异——这样他们才能够进行恰当的回应。

　　步入更高等的学习阶段后，细致的分析变得越来越重要。在科学学习中，你学会将物体分解成部分或更小的单位：元素。在历史学习中，你得关注关键的细节，它们微小却可能改变了历史的走向；在艺术学习中，你会意识到更细微的特征和感官体验的各个方面。

同义动词	相关词汇		智力美德
分辨	不同的	部分	差异化
区分	相反的	元素	剖析
拆开	区别	功能	
分析	异常	复合	
	对比	由两部分组成的	
	然而	边缘	

03

ACTING PLAN

Same but Different

Point out that everyone can divide the room into many different objects, (You could ask, how many?) This is a basic ability. Then say that it is more challenging to recognize differences between two objects that are the same (that is, are given the same name, e.g. book, pen, table, window, etc.) Pairs are invited to choose any such objects and list as many differences as they can. To end, invite each pair to say what is the biggest difference between their chosen objects.

Two but One (Odd One Out)

For this you'll need sets of three objects, pictures or words in which there are similarities between each pair that differentiate them from the other two. The game is to make claims about the odd one out, using the formula, *"Two ... but one ..."*.

For example, with a plane, an eagle and a horse: *"Two are means of transport but one isn't", "Two are natural but one is man-made", "Two can fly but one has to walk."*

03

步法应用

相同但不同

　　每个人都能够很容易地说出组成"房间"的许多种不同物品（教师可以尝试提问：你能在房间中找出多少样物品？），这是一项基本能力。然而，识别两个相同的物体（即被赋予相同名称的物体，例如书、笔、桌子、窗户等）之间的差异则更具挑战性。请学生两两结伴，选择任一种此类物体，尽可能多地列出能想到的差异。最后，邀请每组学生说出他们选择的物品之间最大的差异是什么。

谁出局（单选）

　　列出一组物品（三个为一组），可以使用图片或文字来表示。三个物品之间有多种关联，请指出能够将其中一个物品与其他两个物品区分开来的特点。可以运用这样的句式进行游戏："×××出局，因为这两个……而这一个……"

　　以飞机、鹰和马为例："鹰出局，因为马和飞机是交通工具，而鹰不是"，"飞机出局，因为鹰和马是自然的，而飞机是人造的"，"马出局，因为飞机和鹰会飞，而马只能走。"

Sim/Diff Chart

This is a "compare and contrast exercise", useful in any subject. Two items (objects, characters, events) to be compared will head columns 1 and 3. The middle column is headed "in common", and features or aspects the items have in common are listed here. Features that are different, i.e. unique, for each of the items are listed in the appropriate column. The exercise is best done firstly in pairs, then "twos into fours"—with each pair adding features from the other's lists. This chart is a simpler version of Hyerle's Double Bubble Map.

Applications

Applications
What are the differences between ...?

Design
What are the differences between cantilever and suspension bridges?

PSHE
What are the differences between two branches of government?

English
What are the differences between these two war poems?

Wise Sayings

Every thing is what it is, and not another thing.

——Joseph Butler

Fools ignore complexity. Geniuses remove it.

——Alan Perlis

相同 / 不同图表

　　这是一个"比较和对比练习",适用于任何学科。第一列和第三列中分别书写将要进行比较的两个项目(可以是物品、人物或事件),中间一列的标题是"共同之处",在该列写出两个项目的共有特征。每项独有的特征,则写在一、三列的相应位置。这个练习最好先在两人小组中完成,然后将两人小组合并为四人小组——这样,每组都能从另一组的列表中获得想法的补充。这张图表是海勒双气泡图①的简化版。

学科中的应用

应用

应用的区别是什么?

设计

悬臂桥和悬索桥的区别是什么?

个人、社会及健康教育

这两个政府部门的区别是什么?

英语

这两首战争诗的区别是什么?

名人语录

万物有本然,终不为他者。

——约瑟夫·巴特勒

面对复杂事务,傻瓜选择忽略它,而天才选择克服它。

——艾伦·佩里斯

---

① 大卫·海勒 Thinking Maps 中出现的 8 种图示之一,通常用于对比。——译者注

E

EXPLAIN

解释

黑猩猩

Relating Animal: Chimpanzee

关联理由：聪明而严肃

Relating Reason: Intelligent and serious

01

WHAT IS *EXPLAIN*?

Synonyms	Coaching Questions
Say how	How do we explain this?
Clarify	Could you clarify what you mean?

If you don't understand what is going on, or how things work, you might ask someone to explain—to tell you, as in a story, what causes what, or how things or people relate to each other. A good explanation makes something complicated easier to understand.

01

什么是"解释"？

关键同义词	指导问题
说明如何去做	我们如何对此进行解释？
澄清	你能澄清一下你的意思吗？

    如果你不理解正在发生的事情，或者事物是如何运作的，你可能会请别人解释——像讲故事一样告诉你，是什么因素导致了现在的情形，事物或人是如何相互关联的。好的解释能让复杂的事情变得容易理解。

02

EXPLAIN *EXPLAIN*

Explanations assist our understanding of the world—how it is put together and how it works. The ability to explain well represents a fundamental difference between learning a term's "surface" meaning and "deep" learning, i.e. "deeply" felt meaning. It is the ability to use a term in several contexts and to break it down for others, perhaps employing analogies or metaphors.

We commonly explain by outlining a sequence of events, saying how or why they have happened, giving a context and causes. But we can equally well explain how parts fit together in a whole, as when an architect explains how their building holds together.

If something is badly explained, there can be a call for it to be explained again—that is, for the meaning of the original explanation to be clarified. This may sometimes require a precise definition of just one word. Note that "explaining" (saying how or why something has happened) is different from "justifying" (saying why you have done something).

Synonyms	Alternative Synonyms		Intellectual Virtue
Relate	Because	Motive	Narration
Account for	Story	Law	Definition
Make clear	Narrative	Account	
Define	Cause	(Make) Sense	
	Effect	Process	
	Behaviour	Factor	

02

"解释"步法释义

解释有助于我们更好地理解世界——它是如何构成，以及如何运作的。是否能够对某个术语进行清晰阐释，以显示深度学习（指理解这个术语更深层的含义）和表面学习之间的区别。对某个术语良好的解释力意味着我们能够在不同的情境中去使用它，并可以运用类比或隐喻等方式帮助他人理解它。

我们通常按照事件发生的时间线来对其进行阐释，说明如何发生和为何如此，来为整个事件提供背景和原因。另一种常见的阐释方式是从部分和整体的角度来对事物进行解释，说明各部分如何组合成一个整体，例如建筑师在阐述他们设计的各部分建筑是如何结合在一起的时候。

如果某事解释得不好，对它进行再次的说明有可能是必要的——即需要澄清原始解释的含义。有时可能要对其中的某个词进行精确定义。注意，"解释"（说明某事是如何或为什么发生的）与"证明……有理"（说明你为什么这样做）是不同的。

同义动词	相关词汇		智力美德
讲述	因为	动机	叙述
说明原因	故事	法	定义
明确	叙述	描述	
定义	原因	有意义	
	效果	过程	
	行为	因素	

03

ACTING PLAN

Why Chain

This game involves explaining the causal chains behind events and situations. Start with a problem, large or small, real or imaginary. The student who proffers the problem is the first link in the chain. The next student gives a reason why that happens, the next a reason why that happens and so on.

A: *There are tons of plastic waste in the sea. Why?*
B: *Because people use bottles once and throw them away. Why?*
C: *Because they don't cost much. Why?*
D: *Because they are cheap to make. Why?*
If you make the thinking physical, by getting people to stand next to one another as the chain builds, you can turn your chain into a "Why Tree" with different branches and rival explanations of reasons previously given.

What were the Dominoes?

This can be used to encourage students to explain the sequence of causes of any event, occurrence or phenomenon, in any subject. Draw a row of dominoes on the board, and write the event/occurrence/phenomenon on

03

步法应用

为什么链

这个游戏涉及解释事件和情况背后的因果链。游戏从一个问题开始，这个问题可大可小，可以是真实的，也可以是虚构的。提出问题的学生是这条"为什么链"中的第一环。下一个学生需要对这一问题所描述的现象给出解释并继续提问原因，将这条"为什么链"延续下去。

A：海里有成吨的塑料垃圾。为什么？
B：因为人们用过一次瓶子就扔掉了。为什么？
C：因为它们很便宜。为什么？
D：因为制造成本低廉。为什么？

如果你想让思考变得具象化，可以让人们在建立"为什么链"的过程中，按照链条延展的顺序依次站立，将"为什么链"变成一棵拥有不同的分支"为什么树"，为先前给出的理由，找一找其他解释。

the furthest right. Get students working in pairs to label the remaining dominoes. Each domino has to be the cause of the one to its right, and so you end up with a set of chain-reacting factors, each causing the next.

Applications

Applications
Can you make a Why Chain for ...

PE
Can you make a Why Chain for over-stretching causing injury?

Science
Can you make a Why Chain for a heatwave occurring?

English
Can you make a Why Chain for the character Bottom ending up with the head of a donkey?

Wise Sayings

The way historians explain things is by telling a story.

——Donal Kagan

Whatever cannot be said clearly is probably not being thought clearly either.

——Ludwig Wittgenstein

多米诺骨牌

这个活动适用于任一学科，可鼓励学生解释导致某个事件或现象发生的一系列原因。在黑板上画一排多米诺骨牌，并在最右边写上事件/现象。让学生两两合作，给剩下的多米诺骨牌贴上标签。每一张多米诺骨牌都是导致右边事件/现象发生的原因。这样，最终就将得到一组顺次作用、依序触发的因素。

学科中的应用

应用

你可以为……做一条"为什么链"吗？

体育

请以"过渡拉伸造成伤害。为什么？"开头，做一条"为什么链"。

科学

请以"热浪发生了。为什么？"开头，做一条"为什么链"。

英语

请以"这个角色最后变成了一头驴的头。为什么？"开头，做一条"为什么链"。

名人语录

历史学家通过讲故事来说明历史。

——多纳尔·卡根

如果不能把一件事说明白，很有可能是因为你还没有思考周全。

——路德维希·维特根斯坦

F

FORMULATE

构想

关联动物

海 豚

Relating Animal: Dolphin

关联理由：聪明且顽皮

Relating Reason: Intelligent and playful

01

WHAT IS *FORMULATE* ?

Synonyms	Coaching Questions
Suggest	Does anyone have a suggestion?
Propose	Can you find a way of expressing your idea?

When babies cry or smile, it is obvious they have feelings. At some point, they must have ideas, too, but they cannot formulate, or give form to, those ideas until they have the words to do so—beginning, probably, with "Mama" or "Dadda". After that, they can gradually grow in their ability to express their feelings and to come up with ideas, even theories, of their own.

01

什么是"构想"?

关键同义词	指导问题
建议	有人有建议吗?
提议	你能找到一种表达自己想法的方式吗?

从一定程度上来说,小婴儿也是有想法的。很明显,他们用哭泣或微笑表达自己的感受。但在他们掌握合适的词汇之前,无法将这些感受构建成某个想法 —— 从简单的"妈妈"或"爸爸"开始,他们逐渐增强表达感受的能力,提出想法,甚至是属于他们自己的理论。

02

EXPLAIN *FORMULATE*

Jason Buckley often uses riddles as an icebreaker in workshops. With one class, he was rather baffled by the stony silence. He asked why, unusually, nobody was guessing. A student replied, *"Because we don't know the answer"* But that, of course, is exactly when a guess is needed!

Nurturing a classroom culture that celebrates the expression of ideas, from guesses to theories, suggestions to proposals, encourages learners to "put something out there" and to welcome comment on it. The sooner students realize that you can often build as much on "wrong" answers as on "right" ones, the more they will take risks in response to questions, and the more resilient they will become.

The same culture of risk-taking will help students express their feelings more, and more honestly, in response to you or others, and become more confident in formulating opinions in important matters of judgement.

Synonyms	Alternative Synonyms		Intellectual Virtue
Come up with	Idea	Intuition	Conceptualization
Express	Draft	Proposal	
Invent	Concept	Improvisation	
Speculate	Brainwave	Solution	
	Maybe	Theory	
	Guess	Hypothesis	

02

"构想"步法释义

贾森·巴克利经常使用谜语来作为研讨会的破冰活动。在一次课堂上，学生石沉大海般的沉默让他感到困惑。他奇怪为什么没有人提出猜测。一个学生回答说："因为我们不知道答案。"然而，谜语的答案就是通过猜测得到的呀！

教师应营造一种鼓励学生表达的班级文化，鼓励他们大胆地从猜测到结论、从建议到提案，勇于表达并乐于接纳评论。当学生意识到想法不仅在"正确"的答案上得以建构，也能在"错误"的答案上生发时，他们就会更敢于承担回答问题带来的风险，也将因此变得更加坚韧。

这样的班级文化将帮助学生在回应他人时更充分、更诚实地表达自己的感受，并在重要的决断中更自信地形成观点。

同义动词	相关词汇		智力美德
想出	想法	直觉	概念化
表达	草案	建议	
发明	概念	即兴创作	
推测	灵感	解决方案	
	可能性	理论	
	猜测	假设	

03

ACTING PLAN

Suggest a Title

Display four pictures or photos that could be interpreted in different ways, then invite students to come up privately with titles for them and write them down. Put them in pairs to talk about their suggestions, and then "twos into fours" to agree on one title for each picture. (If they can't easily agree, each person can choose one title for one picture.) Compare and discuss the suggested titles, emphasizing that each one has its merits, and that you want to celebrate creativity, not cultivate competition.

Ideas for Improvement

Give students time, perhaps between classes, to come up with ideas for improving school life or life at large. Then give them 10 minutes in trios to share their ideas and agree on one that they would like to celebrate.

Pair, Pool, Pick, Pitch

This move is useful for any time you ask for suggestions (e.g. to solve a problem). In pairs, students come up with a suggestion (Pair), before forming a four and sharing with others (Pool). The students then decide

03

步法应用

给出一个标题

　　展示四张可以用不同方式解释的图片或照片，请学生独立思考，给每张图片起个标题并写下来，然后两人一组交流自己的想法。之后，将两人组合并为四人组，通过讨论，在小组内就每张图片的标题达成一致意见。（如果最终无法达成一致，允许每人为其中的一张图片选择标题。）比较和讨论大家提出的标题，是为了强调每个标题都有其优点，目的是鼓励创造力，而不是培养竞争。

改进的想法

　　抽一些时间请学生来想一想关于改善学校生活或更广义上的日常生活的想法，课间休息或许是合适的时机。请他们用10分钟，三人一组来分享自己的想法，看看是否能就某个想法达成一致。

which of the two is the better suggestion (Pick) before presenting it to the class (Pitch).

Applications

Applications
Suggest

Geography
... improvements for a cleaner world

Design
... possible or desirable inventions, e.g. a desk that suits everyone

English
... a poem to express your current feeling or mood

Art
... make a doodle or find an interesting picture, then give it a title

Social Science
... ways to reduce the nation's sugar consumption

Wise Saying

Be less curious about people and more curious about ideas.

——Marie Curie

配对、汇聚、挑选、陈述

这个步法适用于任何需要征求建议的时刻（例如解决某个问题时）。学生两两一组（配对）共同提出一个建议；然后将二人小组合并组成四人小组，将建议进行分享（汇聚）；在两个建议中作出抉择，优选更好的那个（挑选）；最后向全班进行建议（陈述）。

学科中的应用

应用
······建议······

地理
为一个更干净的世界而做出改进

设计
可能的或理想的发明，例如适合每个人的桌子

英语
一首表达你当前感觉或心情的诗

艺术
画一个涂鸦或者找一幅有趣的画，然后给它起个名字

社会科学
减少全国糖类消费的方法

名人语录

对人少一点好奇，对想法多一点好奇。

——玛丽·居里

G

GROUP

归类

关联动物

鸭 子

Relating Animal: Duck

关联理由：鸭子常常排队一起走
Relating Reason: Keeping things (ducks) together

01

WHAT IS *GROUP*?

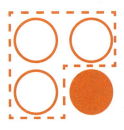

Synonyms	Coaching Questions
Sort	Can we sort these into groups?
Class	How could you class or describe this?

You are often sorting things (objects, people, places) into different groups or kinds. For example, when you tidy things away, you put the same things together. Sometimes the group that a particular thing belongs to is obvious. Other times, it might take some careful thought. Paleontologists have spent years trying to work out which fossils belong to which group or class of dinosaurs.

01

什么是"归类"?

关键同义词	指导问题
分理	我们可以把这些分成几组吗?
分门别类	你会如何对这个进行分类或描述?

　　你经常会将事物(物品、人、地点)分成不同的组别或种类。例如,整理东西时,你会将同样的东西放在一起。有时,某件事物所属的类别是显而易见的;有时,判断它所属的类别则需要你进行仔细地思考。比如古生物学家们花了许多年的时间,试图弄清楚化石所属的恐龙种群或纲目。

02

EXPLAIN *GROUP*

The ability to group things together in our minds is fundamental to human thought, perhaps even instinctive. We recognize something to be a snake, for example, even though we have never met this snake. It is how we make sense of the world and of all the things in it—recognizing that some things have so much in common that they can be regarded as the same sort of thing.

Things can belong to lots of different groups at once. A frying pan can be a container, a potential weapon, an essential purchase for a university student. It can also have a descriptive label, such as (something) "metal/round/shiny/used for cooking". To describe a particular thing is actually to assign it to a group or class of like things. Encouraging learners to class or describe things in different ways challenges them intellectually and makes them think in original, creative ways.

Synonyms	Alternative Synonyms		Intellectual Virtue
Assemble	Same	Belong	Organization
Categorize	Sort	Type	Precision
Label	Kind	Species	
Describe	Set	Category	
	Class	Feature	
	Member	Characteristic	

02

"归类"步法释义

在脑海中将事物进行归类，这是人类思考的基础，甚至可能是一种本能的行为。例如，即使我们从未见过这条蛇，也能认出它是"蛇"。这是我们认识世界及其中所有事物的方式——认识到部分事物有如此多的共同点，因而可以被视为同一类别。

同一事物可以同时分属许多不同的类别。例如，煎锅可以是一个容器、一个潜在的武器、一件大学生的必需品。它还可能有更多描述性的标签，比如"金属的/圆的/闪亮的/用于烹饪的"。实际上，描述某样事物，就是将其归类到某一个与之拥有相似特征的事物类别之中。鼓励学习者以不同的方式对事物进行分类或描述，可以在智力上挑战他们，使他们能用原创的、富有创造力的方式进行思考。

同义动词	相关词汇		智力美德
集合	相同	属于	组织
归类	分理	类别	精度
标签	种类	物种	
描述	套、组	类别、范畴	
	种类、等级	特征	
	成员	特质	

03

ACTING PLAN

Sort It Out!

Take two skipping ropes or hoops and create a Venn diagram on the floor so that you can easily demonstrate the idea of "examples of x go here", "examples of y go there" and "examples of both go in the middle". Label the two rings, for example, land creatures/sea creatures' and give each student an example to sort into the correct space, such as a card with an animal or fish on it. You can increase the difficulty by moving from the factual to the contestable. For example, ask students to sort animals into good pets and bad pets. Then suggest that they sort them into those that are OK to eat and those that are not.

Blockbusters

You'll need to google Blockbusters PPT to bring this 80s game show into your classroom. Two teams compete to complete a path across the screen, which is a tessellation of hexagons labelled with letters. A team selects a hexagon and has to answer a question of the form, "What P is an alkali metal that reacts with water, burning with a lilac flame?" You can get one class to set a grid of questions for another (excellent for revision). Emphasize that each letter labels a group/kind/set/class.

03

步法应用

分类整理！

拿两根跳绳或两个圆环，在地板上创建一个韦恩图，这样就可以用一种简单的方式向大家进行呈现，"x 的例子放在这里""y 的例子放在那里"和"符合两者的例子放在中间"。给两个圆环分别贴上标签，例如，"陆地生物／海洋生物"，给每位学生一个需要进行分类的示例，比如一张动物或鱼类的卡片。你可以通过从事实性类别转向有争议的类别来增加活动的难度，如，让学生将动物分类为"好宠物"和"坏宠物"，或者把动物分类为"可以吃的"和"不可以吃的"。

智力大冲关

你需要搜索"智力大冲关 PPT"，将这个 80 年代的游戏秀带入你的课堂。将全班分成两支队伍，比赛谁先达到终点。屏幕上会有多个标示字母的六边形。每组分别选择其中之一并回答其所代表的问题："P 是一种碱金属，会与水反应，能燃烧出

Sorting Hat

This activity can easily be done with any big set or collection of things (objects, events, artworks, chapters, consequences, symbols—anything you can collect and display). Ask students to sort these things into different groups, like the Sorting Hat sorts pupils into houses at Hogwarts. They may spot categories that wouldn't occur to us. When they've finished, they can inspect other groups' efforts and deduce the categorization chosen.

Applications

Applications
Sort it out!

Music
a range of instruments

Science
a list of elements

History
causes of an event

Wise Sayings

Science is the systematic classification of experience.

——George Henry Lewes

If names be not correct, language is not in accordance with the truth of things?

——Confucius

淡紫色的火焰。P是什么?"你也可以将这个游戏设置成班级间的比赛,让一个班级为另一个班级设置问题网格(非常适合复习),强调每个字母代表一个组/种类/集合/类别。

分院帽

这个活动可以轻松地应用于任何事物集合中(物品、事件、艺术品、章节、后果、符号——任何你能收集和展示的东西)。请学生将这些事物分成不同的组,就像霍格沃茨的分院帽将学生分配到不同的学院一样。他们可能会发现很多意想不到的分类。完成后,请学生相互查看其他小组的分类情况,推断出其他人所选择的分类方式。

学科中的应用

应用
分类游戏!

音乐
一系列的乐器

科学
一系列元素

历史
事件的起因

名人语录

科学是对经验的系统分类。

——乔治·亨利·刘易斯

名不正,则言不顺。

——孔子

H

HEADLINE

标题

关联动物

狮　子

Relating Animal: Lion

关联理由：通过咆哮使信息明晰

Relating Reason: Making clear (roaring) a message

01

WHAT IS *HEADLINE* ?

Synonyms	Coaching Questions
Summarize	How could we headline what X just said?
Distil	Let's summarize/recap the main points...

If you just looked at the headlines of a newspaper, without reading any of the articles, you would still have a pretty good idea of the most important events of the day. Headlines grab attention, are easy to remember and save time. Being able to use a few words to get across the gist of something is helpful in class and in everyday life.

01

什么是“标题”？

关键同义词	指导问题
概括	我们怎样才能用标题概括 X 刚才所说的内容？
提炼	让我们总结 / 回顾一下要点……

　　读报纸时，哪怕你不阅读任何文章的具体内容而仅仅只是浏览标题，你仍然可以对当天最重要的事件有一个比较好的了解。标题吸引注意力，易于记忆，节省时间。能够用几个词概括某事的要点，对课堂理解和日常生活都是很有帮助的。

02

EXPLAIN *HEADLINE*

The essence of all forms of summary is the process of boiling down or distilling something into a more condensed form. Sometimes it can help to Keyword the main ideas, but a headline will typically be a phrase or sentence rather than individual words.

A successful summarizer can process, understand and capture large amounts of information so as to pass on the gist to someone else. It can be very satisfying to come up with a good headline or express an idea with concision.

Summarizing is also the bedrock of Responding or feeding back. A genuine response requires, firstly, an understanding of what has been said. Summarizing is invaluable at every stage of life—from passing on messages to contributing a boardroom discussion.

Synonyms	Alternative Synonyms		Intellectual Virtue
Recap	Heading	Concise	Concision
Abridge	Point	Précis	
Condense	Summary	Bullet points	
Outline	Abstract	Synopsise	
	Digest	Essence	
	Gist	Succinct	
	In a nutshell		

02

"标题"步法释义

所有形式的总结，其本质都是将某事浓缩或提炼得更加精简。有时，关键词化主要观点可能会有所帮助，但标题通常是一个短语或句子，而不是单个词语。

一个成功的总结者能够处理大量信息，理解并捕捉到其中的关键点然后传达给他人。想出一个好的标题或以简洁的方式来表达一个想法是非常棒的。

总结也是回应或反馈的基础。真正的回应首先需要理解对方所说的内容。不论是传递信息还是参与董事会讨论，总结在生活的每个阶段都很有意义。

同义动词	相关词汇		智力美德
扼要重述	标题	简明的	简洁
缩短	重点	梗概	
浓缩	总结	要点	
概述	摘要	概要	
	文摘	本质	
	主旨	简略的	
	简而言之		

03

ACTING PLAN

Seven-Word Summaries

Individually or in pairs, summarize a famous story (book or film) in exactly seven words so others can guess the title. For example, Boy with multi-colored coat dreams himself rich.

Head to Headlines

(Shamelessly adapted from Chris Evans' old breakfast radio show.) A good headline perfectly summarizes in a catchy expression. Once you have covered any material, each student devises a headline for this knowledge. For example, on Photosynthesis: PLANT POWER TURNS LIGHT INTO LIVELINESS! or on Right Angles: TO GET THE RIGHT ANGLE, DON'T CUT CORNERS! Students get into threes and each individual creates a headline. Two announce their headline, with the third student judging which one best sums up the material. The judge's headline is then pitted against the winner's, with the loser deciding between them. Last headline standing wins! Winning headlines from all the trios could then be judged by the teacher or by majority vote.

03

步法应用

七字总结

单独或两两结伴，用7个字（词）概括一个著名故事（书或电影），让其他人来猜测这个故事（书或电影）的标题。例如："彩衣男孩致富梦。（男孩、穿着、五颜六色、外套、梦想、成为、富人。）"

标题赢家

（改编自克里斯·埃文斯以前的早餐广播节目。）一个好的标题能以一种吸引人的表达对事物进行完美概括。在某个知识点的学习结束后，可以请学生为学到的知识设计一个标题。例如，关于"光合作用"：《植物的力量将光转化为活力！》；关于"直角"：《想要得到正确的角度，别走捷径！》，将学生分成三人小组，每个人创造一个标题。其中两名学生先分享自己的标题，由第三名学生判断哪个标题更能概括所学的内容。然后，再将第三名学生的标题与刚才的胜出标题进行对比，请上一轮中落

Hold the Front Page!

(A more formal variation of the above.) When a teacher gives learners a title or learning objective to write at the start of a piece of work, it is the teacher's title for something that doesn't yet belong to the learners. Either instead of or as well as a title, at the end of the work get the learners to write their own, memorable headline in a large font that captures the most important message from what has been learned.

Applications

Applications
Hold the Front Page! Write a headline that captures

Politics
... the Prime Minister's speech to Parliament

Art
... this biography of van Gogh

Geography
... this description of Brazil

Wise Sayings

Much wisdom goes with brevity of speech.

—— Sophocles

If it is possible to cut a word out, always cut it out.

—— George Orwell

败的学生在这两者间进行比较后选择。最后留下的标题获胜！所有在三人小组中胜出的标题，可以由教师或通过班级投票来决定最终的获胜标题。

占据头版！

（这是上一个活动更为正式的一种变体。）学习初始，教师会给学习者一个标题或学习目标，但这是教师为尚未开始学习的学生所拟定的标题，而不是真正属于学习者自己的。在任务完成时，让学习者概括和总结所学知识中最重要的信息，用大字体写下属于他们自己的、令人难忘的标题。

学科中的应用

应用
抓住首页信息！写一个能捕捉到的标题

政治
写一个能够概括首相议会演讲内容的标题

艺术
写一个能够概括凡·高人物小传的标题

地理
为巴西概述写一个标题

名人语录

言简知多。

——索福克勒斯

能砍掉的词，一定要砍掉。

——乔治·奥威尔

I

INFER

推测

关联动物

狗

Relating Animal: Dog

关联理由：通过证据来推测

Relating Reason: Telling from evidence (sniffs)

01

WHAT IS *INFER*?

Synonyms	Coaching Questions
Deduce	Does anyone have a suggestion?
Take from	What might we take from the evidence so far?

To infer is to draw a conclusion, usually from some evidence or from a line of reasoning, a bit like a detective. In class, you might be asked what you can tell from a book or video, but you probably draw lots of conclusions in real life, too, without quite realizing it. If you see frost, for example, you might infer that it is cold and decide to dress warmly, without actually putting your thoughts into words.

01

什么是"推测"?

关键同义词	指导问题
演绎	有人有什么建议吗?
提取	迄今为止,我们可以从证据中得出什么结论?

　　推测就像侦探一样,是从一些证据或一系列推理中得出结论。课堂上,教师可能会问从一本书或一个视频中你所得到的结论;现实生活中,你也可能常常在没有觉察的情况下得出结论。例如,看到霜,你可能会推测出天气很冷,所以决定穿得暖和一点,而实际上你并没有把你的想法用语言表达出来。

02

EXPLAIN *INFER*

Inference takes the form, *"If A (is true), then B(is true)"*, where A is called the "premise", and B is the "conclusion". If B follows necessarily from A—typically because of the way A is defined—the reasoning is known as "formal" or "deductive" logic, and the conclusion is said to be "logical" or "valid". A conclusion is said to be not logical, or "illogical", if it does not follow at all from the premise.

Not all arguments have to be strictly logical. For example, if you see someone with a runny nose, it would be reasonable enough to infer that they have a cold. But they might be suffering from an allergy, not a cold. The conclusion, then, is not a necessary one: it is probable rather than certain. This sort of reasoning is called "informal" or "inductive", and the conclusion is strictly said to be "induction", rather than deduction. Inductive reasoning tends to get better with experience, as we learn what is probable/improbable/impossible; deductive reasoning tends to improve as we learn to distinguish between the meanings of words.

Synonyms	Alternative Synonyms		Intellectual Virtue
Figure out	So	Assumption	Deduction
Conclude	Therefore	Conclusion	Induction
Interpret	If ... then	Implication	
Generalize	Follow(s)	Conjecture	
	Logic(al)	(Not) necessarily	
	Premise	Consistent	

02

"推测"步法释义

推理的形式是："如果 A（是真的），那么 B（也是真的）"，其中 A 被称为"前提"，B 是"结论"。如果 B 必须以 A 为前提，通常是因为 A 的阐述方式——这种推理被称为"形式逻辑"或"演绎逻辑"，而结论被认为是"合乎逻辑的"或"有效的"。如果结论不是由前提得出的，那么这个结论就被认为是不合逻辑的，或称为"非逻辑的"。

并非所有的论证都必须严格合乎逻辑。比如，你看到有人鼻涕直流时推断他们感冒了，这可能是合理的。但这个症状也有可能不是因为感冒，而是过敏。因此，这个结论不是必然的：它是可能的，而不是肯定的。这种推理被称为"非形式的"或"归纳的"，严格来说，结论应该是"归纳"的，而不是"演绎"的。归纳推理随着经验的积累往往会变得更好，因为我们学会了判断什么是可能的 / 不太可能的 / 不可能的；演绎推理水平则随着我们逐渐学会区分词语含义而不断提高。

同义动词	相关词汇		智力美德
想出	所以	假设	演绎
总结	因此	结论	归纳
解读	如果……然后	含义	
推而广之	遵循	猜测	
	逻辑	（不）一定	
	前提	一致的	

03

ACTING PLAN

Detectives

Students wander around a space until you call for them to pair off. Then, in pairs, they look around the room for something to make a statement about, e.g. "One of the lights is not working." One of the pair makes their statement, and the other is given the chance to tell or take something from either the words of the statement or the situation it describes, e.g. "There may be something wrong with the light bulb (or the electric supply)."(The range could be from logically necessary to likely to possible.) On return to their seats, there should be an invitation to celebrate any impressive "detective work" especially inferences of a necessary or probable nature. The exercise can also be done with students sitting in pairs, looking at a photo of an interesting painting (such as *An Experiment on a Bird in the Air Pump*).

Solid, Shaky, Shouldn't

Given a piece of evidence and an inference from it, students indicate with hand signals if they think the conclusion is solid (hand held flat), shaky (hand wobbling) or that the inference doesn't follow (thumbs down), e.g. "This floats, so it can't be made of metal!" Students could be encouraged

03

小小侦探

学生在教室内随意走动，听到教师发出的提示后，和身边小伙伴完成两两结对。两人一组，选择教室里的一件物品并进行描述，其中一个人陈述自己的观点，例如："有一盏灯没有亮。"另一个人从该观点的用词或它描述的情况中获得一些信息并进行推理，例如："它的灯泡（或电源）可能出了问题。"（范围可以从"逻辑上必然"到"可能性较高"再到"可能"。）回到座位上时，邀请大家庆祝任一印象深刻的"侦探工作"，尤其是那些有很高必然性或可能性的推断。这个练习也可以在学生两两结伴观看一幅有趣的画作（如《空气泵实验中的鸟》）的照片时进行。

肯定，不确定，不成立

给学生一个证据及其推论，请他们用手势来表示对该推论的看法。如果他们认为其是坚实肯定的，用平放的手势来表示；认

to assist with the creation of a range of inferences, writing down one that is certain, one that is probable, and one that is only just possible.

What Follows?

Take any central statement or scenario to do with the lesson of the day or week. The task is to come up with as many logical inferences as possible from the statement. For example: "Martin Luther King was a civil rights leader who marched to Selma." What else must be true? (e.g. He visited Alabama, he was not alone; the march had a point ...)

Applications

Applications
What follows?

English
What follows that Scrooge wanted Cratchit to work over Christmas?

ICT
What follows that a computer can beat the world's best chess player?

Math
What follows that 6×7 = 42?

Wise Sayings

From a drop of water a logician could infer the possibility of an Atlantic or a Niagara without having seen or heard of one or the other.

——Sherlock Holmes in A.C.Doyle's *A Study in Scarlet*

We are here and it is now. The way I see it is, after that, everything tends towards guesswork .... I could be wrong. Not being certain is what being a philosopher is all about.

——Didactylos in Terry Pratchett's *Small Cods*

为该推论是不确定的，用摇手表示；认为该推论不成立，则用拇指朝下的手势表示。例如："这个东西是漂浮着的，所以它不可能是金属做的！"可以鼓励学生帮助他们建立一系列推论，写下一个确定的，一个很可能的，一个是仅仅有一点可能的推论。

接下来呢？

任意选择一个与当天或本周课程有关的核心陈述或场景。任务是从陈述中得出尽可能多的逻辑推论。例如"马丁·路德·金是一位民权领袖，曾游行至塞尔玛。"询问学生是否能够从该陈述中推论出其他必然的陈述？（比如：他去过阿拉巴马州，他不是一个人去的；游行是有必要的……）

学科中的应用

应用
接下来呢？

英语
那个吝啬鬼想让克拉奇蒂在圣诞节工作，然后呢？

信息通信技术
电脑打败世界上最好的棋手之后会发生什么？

数学
$6 \times 7 = 42$ 接下来一步是什么？

名人语录

一个逻辑学家不需要亲眼见到或者听到大西洋或尼亚加拉瀑布，他能从一滴水上推测出它有可能存在。

——夏洛克·福尔摩斯　出自柯南·道尔所著《血字的研究》

我们在这里，此时此刻。我的看法是，除此之外，一切都趋向于猜测……我可能是错的。不确定即是一个哲学家的全部。

——迪达哥拉斯　出自特里·普拉切特所著《小鳕鱼》

J

JUSTIFY

证明

啄木鸟

Relating Animal: Woodpecker

关联理由：证明想法（虫子在哪儿）

Relating Reason: Proving your idea (where the bug is)

01

WHAT IS *JUSTIFY*?

Synonyms	Coaching Questions
Give reasons	Who has an argument for (doing) that?
Argue	What reasons could there be for believing this?

To justify means to offer reasons to back up what you believe, or what you have done or want to do, which is important when trying to persuade someone of something! These can't just be any reasons—they need to be ones that would be accepted as relevant and reasonable—so, not just a whim or an unconvincing excuse.

01

什么是"证明"？

关键同义词	指导问题
给出理由	谁有支持（做）那件事的理由？
论证	相信这一点有什么理由？

　　证明意味着为你所信、所行、所欲之事物提供理由，当我们试图说服别人的时候，有力的证明非常重要！"证明"不能只是随意地给出一个理由——需要是相关且合理——才能被接受。所以，不能只是一时的心血来潮或找一个不令人信服的借口。

02

EXPLAIN *JUSTIFY*

JUSTIFY is a natural partner with INFER—almost the other side of the coin. To infer is to draw a conclusion from premises or reasons. To justify, on the other hand, is to have drawn a conclusion (or to have taken an action) and then to provide reasons for doing so. Doing it this way round is actually more common.

Without providing reasons, it's difficult to put a case to someone, let alone persuade them. Furthermore, by valuing reasons, learners will require good ones in order to be persuaded themselves. Habitually giving reasons also encourages learners to reflect on their own positions, double-checking that their views are based on solid reasoning.

It's also important that students understand that while "everyone is entitled to their own opinion", the reasons for some opinions are stronger than for others and affect how much credibility those opinions have.

Synonyms	Alternative Synonyms		Intellectual Virtue
Say why	Reasons	Proposition	Rationality
Excuse	Excuses	Argument	
Prove	Grounds	Proof	
Persuade	Evidence	Principles	
	Position	Rationale	
	Certainty	Valid	

02

"证明"步法释义

"**证明**"是与"**推断**"天然相伴的——几乎可以说是硬币的另一面。"推断"是从前提或理由中得出结论。而"证明"则恰恰相反，是在已经得出结论（或采取了行动）之后，为之提供理由。实际上，这种方式更为常见。

如果不提供理由，你很难向别人陈述案例，更别提说服他们了。重视提供理由，将使学习者对依据的充分性进行自我考量。习惯性地思考自己的行为或结论并提供理由，也为学习者反思所处立场、检查观点是否是基于坚实推理提供了一次复核的机会。

同样重要的是，学生需要理解，虽然每个人都有权选择自己的观点，但某些观点的理由可能比其他的更加有说服力，会影响到观点的可信度。

同义动词	相关词汇		智力美德
说为什么	理由	命题	理性
借口	借口	论证	
证明	依据	证据	
说服	证据	原则	
	立场	理据	
	确定性	有效的	

03

ACTING PLAN

Crazy Reason

This is a justification game in which players provide creative reasons for seemingly preposterous decisions.For example, one person in the pair says, "I always take a hippo to the shops", and the other says, "Of course! It can push people out of the way for you." Then they swap and the other person initiate with an equally preposterous habit.

Why, Why, Why?

Pupils pair up. One person starts by saying something they genuinely believe. Their partner then asks "Why?", and hears the reason. The partner continues to ask "Why?" in response to each reason until the first person can explain no further.

Second Why

This is a good teaching move or mindset. When a learner gives an answer followed by a reason, ask "Why?" a second time. In a philosophical discussion, they'll have to dig deeper inside themselves. If this is a lesson covering content, it requires them to make leaps and connections to other knowledge.

03

步法应用

疯狂的理由

这是一个辩护游戏，在这个游戏中，玩家需要为看似荒谬的决定提供创造性的理由。例如，一个人说，"我一直带着河马去商店"，另一个人说，"当然！河马可以帮你推开人群"。然后他们互换角色，由另一个人提出同样荒谬的决定。

为什么，为什么，为什么？

学生两两配对。由其中一人先开始，说出一件自己深以为然的事情。另一位则负责追问理由："为什么？"听一听伙伴给出的理由。用"为什么？"的句式保持对每一个理由的持续追问，直到第一个人无法给出进一步的解释。

第二个为什么

这是一个很好的教学方法或思维模式。当学习者给出答案并说明理由时，教师可以再次追问："为什么？"在哲思讨论中，多问一个为什么，可以帮助学生更深入地挖掘自己的想法。在具体课程之中，可以要求学生进一步掌握知识并与其他内容进行关联。

Facilitator/Facilitatee

When you have a substantial question to discuss in pairs, put up a PPT slide with:

Can you tell me more?

Can you say why?

So ...? (repeat the question)

Can you give me an example?

How do you mean?

Why is that important?

One of the pair is the facilitator who uses only the questions on the slide to push the facilitatee's thinking deeper. Try to have a second substantial question ready, so that the pair can swap roles.

Applications

Applications
Can you justify your opinion on ...

English
Can you justify your opinion on where the Iron Man came from?

History
Can you justify your opinion on whether King John deserved his nickname?

Science
Can you justify your opinion on whether all energy should be renewable?

Wise Sayings

Prepare your proof before you argue.

——Jewish proverb

Everywhere, authority and tradition have to justify themselves in the face of questions.

——Gustav Heinemann

引导者 / 被引导者

当你有重要的问题需要两人一组进行讨论时，请准备一个 PPT 幻灯片页面，展示以下引导问题：

你能告诉我更多吗？

你能说说为什么吗？

那么……？（重复问题）

你能给我一个例子吗？

你是什么意思？

为什么这很重要？

由一个学生担任引导者，使用幻灯片上的问题来促进另一个学生（被引导者）进行更加深入的思考。教师提前准备好另一个重要的问题，两人小组互换角色，用同样的方式促进另一个学生的深入思考。

学科中的应用

应用
你能证明你对……的观点是合理的吗？

英语
你能证明你对钢铁侠来自哪里的观点是合理的吗？

历史
你能否证明你对"约翰国王是否配得上他的绰号"的观点？

科学
你能否证明你对"是否所有的能源都应该是可再生的"的观点？

名人语录

辩论前准备好证据。

——犹太谚语

不论在哪里，权威和传统都必须在面对质疑时证明自己的合理性。

——古斯塔夫·海涅曼

K

KEYWORD

关键词

关联动物

蜂 鸟

Relating Animal: Hummingbird

关联理由：提取精华

Relating Reason: Picking out essences

01

WHAT IS *KEYWORD* ?

Synonyms	Coaching Questions
Highlight	What are the key words in the lesson so far?
Pinpoint	Can you highlight the key words in the text?

Sometimes it's easy for what is important to get lost in a mass of words. When you have a lot of information to process or remember, it's important to be able to pick out what's essential—the key words that capture the main ideas—or the ones that you're not sure of and need to understand.

01

什么是"关键词"?

关键同义词	指导问题
突出重点	到目前为止,这节课的关键词是什么?
准确定位	你能标出文中的关键词吗?

　　有时候,我们很容易在一大堆文字中漏掉重点。当你有很多信息需要处理或记忆时,辨别出关键的内容是非常重要的。关键词是能够概括主要想法的词语,或者是一些你不确定、需要进一步理解的词语。

02

EXPLAIN *KEYWORD*

Tom remembers setting his students the task of highlighting key information of some text. Expecting sheets to come back with 10% of the words highlighted and 90% blank, he was surprised to see almost everyone had left 10% and highlighted 90%, He realized that, firstly, he assumed his students knew the criteria for what's worth highlighting and, secondly, his pupils were naturally risk-averse to this kind of activity. If in doubt, they highlighted!

Whether or not you can relate to this in your teaching, it's crucial that we regularly give learners the chance to extract the essential ideas from information, so they can proceed to summarize effectively, either for themselves or for others.

Tom also taught a pupil who had an encyclopedic memory for the gory, quirky and humorous bits of history, but had no grasp of the big ideas and concepts involved. Fundamental to making progress is to understand and remember the big ideas, as everything else hangs off them like a clothesline. Let's make sure they're remembering the right stuff!

02

"关键词" 步法释义

汤姆记得，自己曾经让学生标注文章内的重要信息。他预计的状况是，收回作业时，学生应该标记出 10% 左右的文本，而剩下的 90% 则是没有被标注的。但令他惊讶的是，事实上的比例正好相反，学生标出了 90% 的文本，而剩下的 10% 则是未被标注的。他这才意识到，自己预设了学生应该知道什么内容是值得被标记的；此外，学生在本能地规避"标错"的风险：凡是不太能确定的，他们就会标记出来。

无论你是否已经在自己的教学中体会到，我们必须了解的一点是，定期为学习者创造机会是非常重要的。练习从庞杂信息中提取主要观点的能力，这样他们就可以为自己和他人进行有效的总结。

汤姆还教过一个学生，他对历史中血腥、古怪和幽默的零散片段有着百科全书般的记忆，但对历史进程中的重要思想和概念却没有把握。能够取得进步的基础是理解和记住重要概念串成的主线，因为其他一切都像晾衣绳上的衣服一样依附于它们。让我们确保学生记住的是正确的东西！

Synonyms	Alternative Synonyms		Intellectual Virtue
Underline	Important	Core	Acuteness
Spotlight	Main	Fundamental	
Emphasize	Major	Theme	
Essentialize	Basic	Emphases	
	Central	Memorable	
	Essential	Significant	

同义动词	相关词汇		智力美德
在……下画线	重要的	核心	敏锐的
特别关注	主要的	基本的	
强调	重要的	主题	
扼要表达	基本的	重点	
	中心的	值得记住的	
	本质的	有意义的	

03

ACTING PLAN

Keyword Taboo

In pairs, students list five keywords as clues to any well-known film, book or TV program. But it is not allowed ("taboo") for the list to contain any words from the film's title or any of its characters' names. Pairs join another, and each pair in turn reads the first of its keywords. Guesses are invited as to the title, and if they miss the mark then the second word is read out for further guesses and so on. The most successful pairs are those whose titles are guessed with the fourth or fifth keywords. An example is: "Children, animals, adventures, making, ship = Blue Peter?"

Just 3 Words

To review a lesson or a discussion, call for students to write down JUST 3 WORDS to capture the essence of what they have learnt or discussed. These may be separate words, or in the form of a basic sentence. Then have a round of the class, each member giving a word each, checking off any that have already been said so as not to repeat them. If someone forms a three-word sentence, they should offer the whole sentence at once. This device may be used earlier in any lesson after a substantial exchange.

03

步法应用

关键词禁忌

　　学生两两一组，列出五个关键词，另一方依照这些关键词线索，猜测一部知名的电影、书籍或电视节目。关键词里不允许包含电影片名 / 书名 / 节目名中的字或其中任何角色的名字（这是"禁忌"）。然后组与组之间相互进行猜测，每组轮流说出第一个关键词，另一组则对标题进行猜测。如果猜错了或者未猜出，再说出第二个关键词，以此类推。最成功的小组是那些在第四或第五个关键词时被猜中标题的。例如："儿童、动物、冒险、制作、船 = 蓝彼得？"

仅限 3 个词

　　复习一节课或一次讨论，请学生写下 3 个词（只写 3 个词）来概括自己所学到或讨论过程中的精髓。可以使用单独的词，也可以是 3 个词连成的一个简单句。请每个学生轮流发言，给出一个关键词，不可重复（相同的词跳过）。如果某个学生记录

Panning for Gold

Panning for Gold involves rinsing a mix of materials so that the lighter, ordinary, unimportant stuff gets washed away, leaving only the heavier gold behind. Ask students to get rid of the less important words in a text so that only the important words, the "Gold" ones, are still visible. Keywording by elimination!

Applications

Applications
Panning for Gold ...

History
... in this description of Hitler's Germany?

English
... in Ron's letter to Harry?

PE
... in the instructions for trampolining?

Wise Saying

Key words open the doors of understanding.

——Roger Sutcliffe

的关键词是一个完整的简单句，请他直接分享整句话。该活动适用于任何课程，当教师需要结束目前学习或讨论的内容并开始下一主题时，可以选择在课堂上进行这个活动。

淘金

淘金是将混合物进行漂洗，那些重量较轻的杂质会被冲走，最后只留下较重的金子。这个活动要求学生像淘金一样，去除文本中不太重要的词句，只留下最重要的，即"金子"词语。通过淘汰删减来找出文本中的关键词！

学科中的应用

应用
淘金

历史
在描述希特勒掌权德国的文字中找出关键部分

英语
在罗恩给哈利的信里找出关键内容？

体育
……在蹦床的使用说明里找出关键部分？

名人语录

关键词打开了理解之门。

——罗杰·萨特克利夫

L

LISTEN/LOOK

听 / 看

关联动物

马

Relating Animal: Horse

关联理由：超强直觉

Relating Reason: Super aware

01

WHAT IS *LISTEN/LOOK*?

Synonyms	Coaching Questions
Notice	What do you see/hear/sense?
Gather	What have you gathered/found out?

Listening and looking are something we do so constantly we hardly notice the thinking involved. But careful listening and looking are crucial in learning and in understanding the world. Scientists notice things that other people haven't seen. Detectives pick up on small details of what people say in order to get to the truth. Other senses, such as smell, taste and touch, are also important.

01

什么是"听/看"?

关键同义词	指导问题
留意	你看到/听到/感觉到什么？
收集信息	你收集/发现了什么？

　　听和看是我们每天一直在做的事情，我们几乎不会注意到其中蕴含的思考。但是仔细地倾听和观察对于学习和理解世界至关重要。科学家会注意到其他人没有看见的事物；侦探会通过捕捉人们语言中的细节发现真相。其他的感觉，如嗅觉、味觉和触觉，也同样重要。

02

EXPLAIN *LISTEN/LOOK*

One of Art Costa's Habits of Mind is "gathering data with all the senses".

This is a vital Thinking Move or habit, because if we are not careful, we can get lost in a world of ideas. Reminding ourselves to notice what is going on around us, or within us, keeps us in touch with the real world.

But it is not just raw data that we gather through our senses. When we hear words (or read them, by sight or touch) we are able make sense of information or instruction that others are trying to communicate to us. And some information is so valuable to us that we actively seek it out, in books or elsewhere. In other words, Listen/Look can be as advanced a move as researching.

Finally, numerous studies attest to the benefits of attending to our inner world and training ourselves to keep calm or mindful in a world of rolling news, notifications and (social) networking. There is also a sense in which we may listen to our intuitions or gut feelings—which can be as important a move as listening to reasons and arguments, especially when we come to Weigh up.

Synonyms	Alternative Synonyms		Intellectual Virtue
Perceive	Senses	Environment	Attention
Observe	Aware	Information	Comprehension
Apprehend	Alert	Message	
Make sense	Perception	Communication	
	Sensation	Mindful	
	Sensitive	Introspection	
	Observant		

02

"听 / 看" 步法释义

阿特·科斯塔的思维习惯之一是"用所有感官收集信息"。这是一个非常重要的思考动作或思考习惯。因为稍不留神，我们可能就会迷失在想法的世界里，常常提醒自己注意周围或自己内心正在发生的事情，与现实世界保持联系。

但我们通过感官收集的不仅仅是原始数据。在听到（或通过视觉、触觉阅读）单词时，我们能够理解其他人试图传达的信息或指令。有些信息非常有价值，所以我们可能会积极地通过书籍或其他渠道进行更进一步的收集。换句话说，"听 / 看"也可以是像"研究"一样的高阶步法。

最后，众多研究证明，在一个充斥着滚动新闻、通知和（社交）网络的世界中，多关注自己的内心世界，训练自己的专注度，学会保持冷静是大有裨益的。这里有一点要注意，我们可能也会听从自己的直觉或直观感受——它们有时和倾听理由及论证一样重要，尤其是当我们需要进行权衡的时候。

同义动词	相关词汇		智力美德
感知	感官 / 感觉	环境	注意
观察	意识到	信息	理解
领会	警告	消息	
使……成立	感知	沟通	
	感觉	觉察	
	敏感的	内省	
	善于观察的		

03

ACTING PLAN

Attention!

There is a world of difference between randomly looking around and deliberately looking for something, with senses attuned to a particular objective. Ask the class to look around the room. Then give the name of a color and ask students to look around again. What did they notice the second time but not the first? It's remarkable how even tiny details stand out.

Resting Receptors

Ask students to sit comfortably and close their eyes. Explain that they are to use their senses to follow your instructions, in silence, in their head. How many different sounds can they hear? What are their aches and pains? Are they sitting comfortably? Does their last meal still linger on their taste-buds? How long does each breath last? After a few minutes, let them share their answers with a partner and then with the group.

Did I Miss Something?

Player A speaks to Player B for 90 seconds on a true story or the current topic of study. B must then verbally recall to A as much as he or she

03

步法应用

注意！

　　"随意地四处看看"和"有意识地运用感官寻找某样东西"之间，存在着天壤之别。请大家先环视教室，随意地四处看看。然后给出一种颜色，请大家再次环视教室时有意识地关注该颜色。第二次观察时的新发现是什么？带着意识时运用感官时，微小的细节也会变得瞩目起来。

感官休息

　　请学生找一个舒适的姿势坐好，闭上眼睛，跟随教师发出的指令问题，安静地进行感受：你能听到多少种不同的声音？哪里感觉到了疼痛或不适？你坐得舒服吗？你的味蕾还能感受到最后一餐所吃食物的味道吗？每一次呼吸持续多长时间？几分钟后，请学生先与身边的同伴分享自己的答案，然后在整个班级中进行分享。

remembers. When they have recalled all they can, B asks, "Did I miss something?" Player A reminds them anything they've missed. If it's about the topic under study, Player A can then also ask, "Did I miss something?" with B supplying missing details.

Applications

Applications
Did I miss something?
An excellent revision tool for storylines and processes

English
What happens in *Romeo and Juliet*?

Science
What happens in the water cycle?

Wise Saying

You see, but you do not observe. The distinction is clear.

——Sherlock Holmes in A. C. Doyle's
A Scandal in Bohemia

我漏掉什么了吗?

玩家 A 向玩家 B 讲述一个真实的故事或是一个当前正在学习的主题,叙述时间为 90 秒。叙述结束后,由 B 口头回忆自己记得的内容。完成后(回忆出所有能想起的内容),B 问:"我遗漏了什么吗?"A 进行提醒。如果是关于一个大家目前正在学习的主题,A 也可以发问:"我有没有什么遗漏的?"B 可以补充遗漏的细节。

学科中的应用

应用

我漏掉什么了吗?
对故事线和流程来说,这是一个优秀的复习工具

英语

《罗密欧与朱丽叶》里发生了什么?

科学

水循环中发生了什么?

名人语录

你在看,却没有在观察,这两者是完全不同的。

——夏洛克·福尔摩斯 出自柯南·道尔所著《波希米亚丑闻》

M
MAINTAIN

保持

公 牛

Relating Animal: Bull

关联理由：明确坚持自己的立场

Relating Reason: Taking a strong stand

01

WAHT IS *MAINTAIN*?

Synonyms	Coaching Questions
Believe	Who agrees with that statement?
Affirm	Could someone take a position on this?

To maintain a belief is to hold that it is true or valuable. It is not just having a belief but consciously committing to it. We may do this privately, i.e. without telling anyone, but if we do make our belief public, someone may Test or doubt it. In that case, we would probably need to Justify it. Or maybe it is someone else's belief that is being doubted but we agree with it. Then if we declare our support for their position, that would be another form of maintaining.

01

什么是"保持"？

关键同义词	指导问题
相信	谁同意这个观点？
断言	有人能对这个问题表明立场吗？

　　坚持某个观点意味着你坚定地认为它是真实的或有价值的。"保持"不仅止于持有观点，更在于你在有意识地维护它。有时候，你可能在心里"保持"了某个观点，但并没有告诉其他人。可一旦公开表达了自己的观点，你就必须做好受到他人测试或质疑的准备，此时，"保持"观点意味着需要去证明它。或者是一个由别人提出的被质疑的观点却被你认可。当你声明自己支持该观点时，这就是"保持"的另一种形式。

02

EXPLAIN *MAINTAIN*

There are pastoral as well as academic benefits to developing students' ability to maintain. Pastorally, we don't want them to "follow their friends", but think for themselves and act with integrity. We should therefore encourage them to maintain or assert their point of view, especially if it might differ from their friends' or the majority's. We should also encourage them to praise and support others who may be in a minority—a move that's reportedly becoming rarer in the workplace.

Academically, we want students to have the confidence to take a position and be ready to defend or justify it. Without this intellectual challenge, we run the risk of developing compliant students who are neither critical nor creative. It is also important that they learn to recognize when someone else is taking a position—which is not always obvious—and then to position themselves in response.

Synonyms	Alternative Synonyms		Intellectual Virtue
Agree	Belief	Committed	Conviction
Support	Claim	Value	
Hold	True	Principle	
Assert	Reality	Worldview	
	Position	Axiom	
	Point of view		

02

"保持" 步法释义

　　培养保持（观点）的能力对学生的生活和学业发展都有好处。生活中，我们不希望他们盲从朋友，而要独立思考，诚实行事。因此，应该鼓励他们坚持自己的观点，特别是当观点与他们的朋友或大多数人的观点不同时。我们还应该鼓励他们赞扬和支持少数群体——据报道，这种做法在职场上正变得越来越少见。

　　在学术上，我们希望学生自信地表明立场，并准备好为自己的选择进行证明和辩护。如果缺乏这种智识挑战，我们就面临着培养出顺从但缺失创新能力和批判意识学生的风险。对学生来说，学会识别他人（可能并不总是显而易见）的立场也很重要，他们需要在此基础上进行思考和选择，然后给予回应。

同义动词	相关词汇		智力美德
同意	信仰	承诺	确信
支持	声称	价值	
持有	真的	原则	
断言	现实	世界观	
	位置 / 立场	公理	
	观点		

03

ACTING PLAN

True or False?

A simple game to help students be clear when someone is making a "truth claim". You or the pupils can make up statements for a quiz or find examples on the Internet.

Change Places If (Contestable Statements)

Put students in a circle or horseshoe. Stand in the middle, with no spare seats available, and say, "Change places if you believe that girls are better students than boys" or some such contestable statement. Ask students holding the minority belief to stand at a distance from each other in the middle of the circle. Invite those who remain seated, one at a time, to stand with someone in the middle. Make sure everyone standing has a pair before everyone else joins a group. Tell students to justify their position in their small groups before they sit down again.

Emphasize that the belief you put forward was contestable—a "Value judgement". Ask students to agree, in pairs, on two value judgements they think others will disagree with. Invite a volunteer to swap places with you and complete "Change places if you believe ..." with a contestable statement of their own, making sure they go to sit in someone else's

03

步法应用

真或假？

这是一个简单的游戏，旨在帮助学生更清晰地评判他人所下的断言。由教师或学生提出或者从网络上找一些陈述性的观点，练习评判其真假。

如果……，请交换位置（有争议的声明）

让学生围坐成一个圈或马蹄形，确保没有多余的座位。发指令的人站在中间，发表指令："如果你认为女孩比男孩更优秀，请交换座位"，指令可以是类似的带有争议性的观点。交换好后，教师比较持两方观点学生的人数，请人数较少的一方站到圈中间来，彼此保持一定的间隔。然后邀请坐着的学生依次站到圈中人身边，确保圈中每个站立的人都有一个伙伴，再分配成为人数更多的小组。请学生在听到再次坐下的指令之前，在小组中为自己的立场进行辩护和证明。

要强调你的指令中涉及的观点是有争议的——是一种"价值判断"。让学生两两一组，思考并准备好两个指令：指令中包含其他人可能会不同意、具有争议性的观点。邀请一名学生代替教师担任发出指令的角色（教师进入游戏，使座位数比参与人数少 1），

place. A new student should be left to offer a contestable statement. Let everyone have fun changing places, but pause for discussion if the belief is particularly contestable.

Bracketing Beliefs

Give students a text, from a text book or newspaper, and ask them to bracket off each assertion or truth claim. You can model this, showing that some sentences contain more than one assertion, and that some assertions are not truth claims but opinions. If the latter, they should also be underlined. (Some sentences will be instructions not assertions.) When students have finished, ask them to count the pairs of brackets. Do they all agree on the number? Are some assertions more questionable than others?

Applications

Applications
Change places if you think ...

English
... "devious" is the most accurate word to describe Fagan

Music
... this piece of music fits into the Baroque genre

History
... numbers were the biggest factor in William conquering Norman England?

Wise Sayings

I maintain that nothing useful and lasting can emerge from violence.

——Shirin Ebadi

If you don't stand for something you will fall for anything.

——Gordon A. Eadie

用"如果你认为……，请交换位置。"的句式来发出指令，确保符合指令的人进行了位置交换。每一轮游戏后，都会有一个学生因为没有入座而成为下一个发出指令的人。让大家开心地享受游戏，但如果出现了特别有争议的观点，可以就此暂停进行讨论。

观点括号

给学生一段来自教科书或报纸的文本，请他们用括号标示出其中的主张或真理。教师可以给出示范，标出一些包含多个观点的句子，以及一些不是真理而是主张的观点。除了用括号标示之外，可以将后者用下画线标示出来，以示区分。（句子中的一些部分可能只是"说明"而非主张。）完成后，请学生数一数一共找到了几对括号。大家都同意这个数字吗？是否有些观点比其他的更存在争议呢？

学科中的应用

应用
如果你认为……，请交换位置

英语
"狡猾"是形容费根最准确的词

音乐
这首乐曲属于巴洛克风格

历史
数字是威廉征服诺曼英格兰的最大因素？

名人语录

我认为暴力不会产生任何有用和持久的结果。

——希尔琳·艾芭迪

如果没有信念，你将会轻易相信任何东西。

——戈登·伊迪

N

NEGATE

否定

关联动物

蛇

Relating Animal: Snake

关联理由：发出嘶嘶声表示拒绝

Relating Reason: Saying no (hissing)

01

WHAT IS *NEGATE* ?

Synonyms	Coaching Questions
Disagree	Does anyone disbelieve what X says?
Oppose	Does anyone disagree with Y's position?

This move is the opposite of Maintain. Here, we express our disagreement with someone, perhaps denying the truth or value of their claim, or the relevance or validity of their argument. People can shy away from disagreeing with others, but disagreeing with someone is a way of respecting them and taking their ideas seriously, especially when you make clear that it's the idea you are disagreeing with rather than the person.

01

什么是"否定"?

关键同义词	指导问题
不同意	有人不相信 X 所说的吗?
反对	有人不同意 Y 的立场吗?

　　否定与保持相反。否定指不同意某人的意见,在这里,不同意可能表现在否定其主张的真实性或价值,抑或是否认其论点的相关性或有效性。人们可能会倾向避免与他人发生意见不合,但事实上,发表不同的想法,其实恰恰是一种尊重他人、认真对待他人想法的方式,尤其是当你明确表达了自己不同意的是对方的想法,而非针对对方本人时。

02

EXPLAIN *NEGATE*

At its most effective, negation is done by producing a counter-example or even a whole counter-argument. But it may also be done by simply contradicting or by offering an alternative account or perspective. It's arguably the basis of critical thinking.

Helping students feel comfortable disagreeing with one another and not feel personally under attack when their ideas are challenged, is crucial for open discussion. Artificial debates with students assigned to one side or another can be helpful, though skill in argument should never be valued or cultivated more than sincerity. Terms like those in the left column ensure that discussions are focused on a clash of ideas, rather than of personality.

Synonyms	Alternative Synonyms		Intellectual Virtue
Disagree	Disagree	Counter	Opposition
Object	Wrong	Contradiction	
Deny	False	Denial	
Dispute	Negative	Rebuttal	
	Opposite	Challenge	
	Contrary	Antithesis	

02

"否定" 步法释义

最有效的否定是通过提出一个反例甚至整个反论证来完成的。但也可以通过简单的反驳或提供替代性的解释或观点来实现。可以说，否定是批判性思维的基础。

在公开讨论中，帮助学生建立对异见的接纳，接受彼此之间可以有不同的想法。当自己的观点受到挑战时，他们不会感觉受到了攻击，这一点至关重要。和学生练习辩论可能会有所帮助，但对辩论的技巧的重视永远不应该超越对真诚之心的珍惜。使用类似左栏中的术语，提示我们讨论的重点应该是思想的差异，而不是个性的冲突。

同义动词	相关词汇		智力美德
不同意	不同意	反驳	反对
反对	错误的	矛盾	
否认	假的	否认	
争议	负面的	辩驳	
	反面的	挑战	
	相反的	反论	

03

ACTING PLAN

But What If ... ?

This paired game is a good way to practice the focused turn-taking involved in negation—and reassertion, which is sort of negation of negation! It's easiest explained with an example:

A: What if you wanted to know the time?

B: I'd look at my phone.

A: But what if your phone was out of battery?

B: I'd charge it up.

A: But what if you'd lost your charger?

B: I'd use someone else's.

A: But what if there was a power cut?

B: I'd use the clock in the school hall.

A: But what if the power cut lasted so long, all the batteries had run out?

B: I'd see when the sun came up.

What if you wanted to meet the Queen?

What if you wanted to go to sleep?

What if you wanted to eat some ice cream?

03

但如果……?

这个双人游戏聚焦"否定"步法，练习"否定"并鼓励再次思考，有点像对否定的否定！用一个例子就会很好理解：

A：如果你想知道时间怎么办？

B：我会看手机。

A：但如果你的手机没电了怎么办？

B：我会给它充电。

A：但如果你丢了充电器怎么办？

B：我会用别人的。

A：但如果停电了怎么办？

B：我会用学校大厅的钟。

A：但如果停电持续很久，所有的电池都耗尽了怎么办？

B：我会看太阳什么时候升起。

如果你想见女王怎么办？

如果你想睡觉怎么办？

如果你想吃点冰激凌怎么办？

Disagreement Duet

Pair up with a friend. Find as many things as possible that you disagree about in 147 seconds. Afterwards, choose one and try to persuade the other to change their mind.

Brain Stand

During a discussion, get everyone to do a Brain Stand, swapping sides and arguing the opposite to what they believe. Standing physically on opposing sides helps to turn the discussion into more of a game.

Applications

Applications
Who has a counterargument / counter-example to ...

Philosophy
... what Annie just said?

Science
... the view that science can explain everything?

Math
... the working out of this equation?

Wise Saying

Without contraries there is no progression.

——William Blake

分歧二重奏

两两一组，在 147 秒内尽可能多地找出双方存在分歧的事情。然后选择其中一件，尝试说服对方改变看法。

立场互换

在讨论过程中，让每个人都进行一次立场互换，尝试为与自己观点相反的一方进行辩论，以此来锻炼立足于不同立场思考问题的能力。可以使用具象的形式，请学生在立场互换时，在空间位置上也进行交换，让活动更具游戏性。

学科中的应用

应用

谁对……有反论 / 反例？

哲学

谁对安妮刚才说的有反论 / 反例？

科学

谁能对"科学可以解释一切"的观点提出反论 / 反例？

数学

谁对这个方程的解有反对意见？

名人语录

没有对立就没有进步。

——威廉·布莱克

O

ORDER

排序

海　狸

Relating Animal: Beavers

关联理由：以有序的方式做某事（建造水坝）

Relating Reason: Building things (dams) in an orderly way

01

WHAT IS *ORDER*?

Synonyms	Coaching Questions
Sequence	Let's put these things into an order ...
Arrange	Do we need a plan for this?

Things can be put into order in time (e.g. events into a sequence), or in space (places mapped out, furniture arranged), or in any other order that makes things easier to think about and deal with—from largest to smallest, most certain to most doubtful, best to worst. You could even order ways of ordering things from the most to the least useful.

01

什么是"排序"?

关键同义词	指导问题
列序	让我们将这些事物按顺序排列……
安排	我们需要为此制定一个计划吗?

　　事物可以按时间顺序排列(例如,给发生的事件排列顺序),或按空间顺序排列(如地点绘制、家具布置),或按其他能使事物更易于被思考和处理的顺序进行排列——从最大到最小,从最确定到最可疑,从最好到最差,甚至可以是从最有用到最无用。

02

EXPLAIN *ORDER*

Students often don't get as much practice in this move as they might, as we tend to do a lot of the organizing and planning for them. We have to be willing to have some chaos at the start for them to bring some order to!

As well as arranging things in time and space, ordering can be about ranking things in order of magnitude, importance, usefulness, beauty, impact—or many other concepts, objective or contestable. Ranking or prioritizing helps students to be resourceful and keep on top of their own work and life-loads.

Having everything ordered for you makes for an easy life, but it's also disempowering. We can help students become self-organizers by eschewing micro-management—instead, encouraging them to set their own goals and allowing them to choose the means and plan the steps to achieving them.

Synonyms	Alternative Synonyms		Intellectual Virtue
Line up	First/Next	System	Orderliness
Timetable	Step/Stage	Procedure	
Layout	Series	Line	
Map	Timetable	Layout	
	Plan	Map	
	Method	Coordinates	

02

"排序"步法释义

学生通常可能没有机会得到更多这方面的练习，因为成人总是替他们做了很多组织和计划的工作。我们需要接受开始时的混乱，给孩子们一些尝试排序的机会。

除了根据时间和空间排序之外，也可以按照数量级、重要性、有用性、美观性、影响力以及其他更多的概念（可以是客观的也可以是有争议的）来对事物进行排序。给出排名或排列出优先级可以使学生能够更为机敏地把握自己的学业和生活。

有人为你安排好一切确实使生活变得更轻松了，但这也会让人失去自我规划的能力。尽量避免微观管理，帮助学生成为自我组织者——鼓励他们设定自己的目标，并允许他们自己规划实现这些目标的步骤和方法。

同义动词	相关词汇		智力美德
列队	第一个／下一个	系统	有序
为……安排时间	步骤／阶段	程序	
布局	系列	路线	
绘制地图	时间表	布局	
	计划	地图	
	方法	坐标	

03

ACTING PLAN

How to Make ...?

Place two slices of bread, jam, butter and a butter-knife on your desk. Ask pairs or threes to produce a sequence of instructions to enable a robot to make a jam sandwich with them. After 5—10 minutes, let students read their instructions while you follow their commands as a non-thinking robot would, to inevitable hilarity! Other simple activities that require methodical steps are making a cup of tea or cleaning your teeth. Students could present each other with similar challenges.

Order, Order, Order!

For any "group of things" you've encountered in the curriculum, get students to write each on a piece of paper. Ask them to put them into an order (of their own choosing) and when done, invite a neighboring group to guess what the thinking behind the order is. This can be repeated with different orders and groups challenging each other.

03

步法应用

如何做?

将两片面包、果酱、黄油和一把黄油刀放在桌子上。请学生 2—3 人一组,为"果酱三明治机器人"制定一系列操作指令。5—10 分钟后,由教师扮演这个不会思考的机器人,让一组学生读出自己的指令,教师则完全听从指令操作。这一定很好玩!还可以以这种方式,为其他类似的简单活动制定步骤,如"泡茶"或"刷牙"。学生也可以自创内容来相互挑战。

排序,排序,排序!

随机选择课堂中出现的"一组事物",请学生在纸上将它们写下来,并按照某种方式排列(排列方式由学生自己选择)。完成后,请旁边组的学生来猜测该排序背后的想法。可以在不同的小组或不同的排序方式中重复该挑战。

First Moves First

Before any given project or piece of work, ask learners what move they think needs to be made first. Once one is agreed, ask someone to stand in a space to represent that move. Ask what would be needed next, and the person representing it puts their hand on the shoulder of the first. Continue the process, creating a planned sequence of moves for the task ahead.

Applications

Applications
Put these into an order

Music
Brass instruments

History
Victorian inventions

Science
Features of the rainforest

Wise Saying

Good order is the foundation of all things.

———Edmund Burke

行动序列

在进行某个项目或任务之前，询问学生第一步需要做什么。达成一致后，请一个学生站在空间里代表"第一步"。然后询问下一步需要做什么，请另一个学生代表"第二步"，并把手放在第一个人的肩膀上。继续这个过程，为接下来需要进行的项目 / 任务创建一个有计划的行动序列。

学科中的应用

应用

把这些按顺序整理好……

音乐

铜管乐器

历史

维多利亚时代的发明

科学

热带雨林的特征

名人语录

良好的秩序是一切的基础。

——埃德蒙·伯克

P

PICTURE

描绘

兰花螳螂

Relating Animal: Orchid mantis

关联理由：善于想象（假装是一朵花）

Relating Reason: Imagining (pretending to be a flower)

01

WHAT IS *PICTURE* ?

Synonyms	Coaching Questions
Imagine	Let's picture the scene ...
Put yourself	Put yourself in the scene / shoes of ...

Sometimes you can make better sense of things or ideas by trying hard to picture them, as in a movie. You can even put yourself into the picture—imagining that you are in somebody else's shoes, trying to see what they would see and feel what they would feel.

01

什么是"描绘"？

关键同义词	指导问题
想象	让我们想象一下这个场景……
设身处地	将自己置于……的场景 / 立场中……

　　像为电影制作一帧帧图像那样地将事物或想法在脑海中描绘出来，你或许可以通过这个过程，尝试着更好地理解它们。你甚至可以将自己置入其中——想象自己处于别人的立场，试着去观看他们会看到的东西，感受他们所产生的情感。

02

EXPLAIN *PICTURE*

To picture is to see something in the mind's eye, and this move has many uses. One is to ensure that your thinking is precise and concrete. If you cannot picture something with clarity and detail, you may overlook a significant aspect of it. Before you buy a sofa, you do well to visualize (as well as Size) your living room.

Another use is "envisioning" the future, a vital aspect of creativity, which involves imagining things other than as they are. Visualizing your future self is also a common technique in sports psychology and can help raise aspirations. It's also potentially a rich experience for imaginative immersion in a topic area, particularly using prior learning to visualize the world of a text or historical era.

Yet another important use is that of imagining yourself in another's shoes. It is a cognitive move, certainly, but it could be seen as a moral move, too—one that most moves us to feel empathy and display compassion towards others (as well as sharing in their happiness).

02

"描绘"步法释义

想象就是用心灵之眼去看东西，在脑海中描绘出某物的样子。**"描绘"**是一个用途广泛的思考步法，其一它确保了想法的精确和具体。如果不能清晰而详细地描绘某件事，意味着你可能忽略了它的某个重要维度。就像在买沙发之前，你最好先在脑海中将客厅可视化（还有度量）一下。

另一个用法是"展望"未来，这是创造力的一个重要方面，它包括超越事物现有的样态去进行想象。在运动心理学中，想象未来的自己是一种常见的技巧，可以帮助运动员提高自我预期。基于某个主题来进行沉浸式想象也可能会给学生带来丰富的体验，特别是请他们使用先前的学习经验，将某段文本或某个历史时代进行可视化时。

另一个重要的用法是想象自己站在别人的立场上。当然，这是一种认知层面的行为，但它也可以被视为一种道德层面的行为——一种最能让我们对他人产生同理心、表现出同情（以及分享他们的幸福）的行为。

Synonyms	Alternative Synonyms		Intellectual Virtue
Visualize	Scene/Scenario	Projection	Imagination
Conjure up	Mind's eye	Make believe	
Pretend	Representation	Daydream	
Immerse yourself	Image	Immersion	
	Vision	In their shoes/place	
	Model	Empathy	

同义动词	相关词汇		智力美德
可视化	场景 / 情境	投影	想象力
在脑海中浮现	心灵之眼	使相信	
假装	表现	白日梦	
让自己沉浸其中	图像	专心	
	愿景	设身处地	
	模型	同理心	

03

ACTING PLAN

What's in Your Picture?

Practice firstly with the whole group, asking everyone to picture an object, such as car, in their mind's eye. Emphasize that you want them to visualize their object as fully and clearly as they can. Then ask them questions of detail, such as "What color is the car in your picture?" or "Does it have an aerial?". (It doesn't matter if the student has not visualized any particular detail. The richness of people's visual pictures and capacity to hold them in mind varies widely: it's an interesting internal difference.) Then students play in pairs. One visualizes an object or scene (either one known to them, or an imaginary/"typical" one) and the other asks questions of detail and tries to re-imagine or copy their partner's image. After a while, the partners reverse their roles. This game is particularly valuable in History and Literature.

Mind's I

Using a picture that includes a human character within the context under study, ask students to look at the picture and fix its details in their memory. Then ask them to close their eyes and imagine they are the person in the picture. What can they see? What can they hear? What do they feel? What are they thinking? What one line of words are they saying

03

你的画面里有什么？

首先和全班学生共同练习，让每个人尝试在脑海中使用"心灵之眼"描绘一个物体，比如汽车。要求他们尽可能完整、清晰地进行想象。教师可以提问一些关于细节的问题，比如"你想象中的汽车是什么颜色的？"或"它有天线吗？"（如果学生未能视觉化一些特定的细节也没关系，人们形成视觉图像的丰富度以及对视觉图像的记忆力差异很大：这是一种有趣的内在差异。）之后，请学生在 2 人小组中继续进行这个练习，一个人想象某件物体或场景（可以是他们熟悉的，也可以是想象中的或者"典型的"），另一个人来询问细节问题，并尝试重新想象或复述同伴的图像。游戏进行一段时间后，两两交换角色。这个游戏在历史和文学学科中特别有价值。

心灵的我

提供一些包含人物角色的图片，内容可以和近期学习内容相关，请学生观察图片并记住其细节。然后让他们闭上眼睛，想象自己是图片中的那个人。他们能看到什么？他们能听到什

to themselves? They can then return to their classroom selves and share what their lines of internal monologue were in pairs and onwards to the class.

Picture Perfect?

Students sit back to back, each with two pieces of blank A4, on one of which each does a drawing (not too simple, not too complex). Each in turn describes their drawing to their partner, who tries to replicate it on their blank sheet. The descriptions should not include the names of any drawn items but may name shapes or similar items. Pictures are compared at the end.

Applications

Applications
In your mind, picture ...

History
In your mind, picture Rosa Parks as she took her seat on the bus

Art
In your mind, picture van Gogh as he painted the cafe at night

Math
In your mind, picture what a 2D world is like

Wise Sayings

Imagination is more important than knowledge. Knowledge is limited. Imagination embraces the entire world.

——Albert Einstein

Formulate and stamp indelibly on your mind a mental picture of yourself as succeeding. Hold this picture tenaciously. Never permit it to fade. Your mind will seek to develop the picture ... Do not build up obstacles in your imagination.

——Norman Vincent Peale

么？他们能感受到什么？他们在想什么？他们对自己说了些什么？然后请学生回归现实——自己的课堂身份，与伙伴分享自己的内心独白，然后再分享给全班。

画面完美吗？

请学生背靠背坐好，每人拿着两张空白 A4 纸，同学 A 先在纸上画一幅图（不要太简单，也不要太复杂）。画完后将自己的画描述给同学 B，同学 B 则根据听到的内容，在另一张白纸上复制 A 同学的画。描述中不能包含任何所绘物品的名称，但可以说明物品的形状或描述类似的物品。最后比较两幅图。完成后可以交换角色。

学科中的应用

应用
在你的脑海中描绘……

历史
在你的脑海中描绘罗莎·帕克斯在公共汽车上坐下

艺术
在你的脑海中描绘凡·高在晚上画咖啡馆的样子

数学
在你的脑海中描绘 2D 世界

名人语录

想象力比知识更重要。知识是有限的，而想象力却包含整个世界。

——阿尔伯特·爱因斯坦

在脑海中形成并镌刻一幅自己成功的画面。持续保持，永远不要让它褪色。你的大脑会试图展开这个画面……不要在想象中为自己设置障碍。

——诺曼·文森特·皮尔

Q
QUESTION

提问

关联动物

牛 虻

Relating Animal: Gadfly

关联理由：持续的好奇心

Relating Reason: Persistent curiosity

01

WHAT IS *QUESTION*?

Synonyms	Coaching Questions
Ask	What questions could we ask about this?
Wonder	Let's get into wonder mode …

In school, you spend a lot of time answering teachers' questions. But to enjoy learning independently, you need to ask questions yourselves. Questioning is important for practical things, such as getting information and advice, but also for the pleasure of being curious. (Could scientists ever genetically engineer a unicorn?) It also helps you work out it if the things other people tell you are actually true! (Is playing video games really bad for your brain?)

01

什么是"提问"?

关键同义词	指导问题
询问	对此,我们能提出哪些问题?
好奇	让我们进入好奇模式……

在学校,你在回答教师提出的问题上花了很多时间,但如果想要享受独立学习的乐趣,你需要尝试自己来提问。提问是实际的,它对解决实务类的事情来说很重要,比如当你希望获取信息和建议时;提问也可以是天马行空的,它能让我们保持好奇的快乐。(比如科学家能对独角兽进行基因改造吗?)它还可以帮你弄清楚别人告诉你的事情是否属实!(比如玩电子游戏真的对大脑有害吗?)

02

EXPLAIN *QUESTION*

Questioning is a skill at which young students excel but which can fade as childhood progresses into adolescence. Learners need to develop the will, as well as the skills, to inquire. This comes through practice: developing questioning as a regular habit, but also celebrating spontaneous questioning.

Sometimes a teacher might respond with an encouragement (or better still, some time) to do some private research. At other times, with a probing question of her own, the teacher encourages and empowers pupils to think things through for themselves.

Questioning in the classroom is notoriously dominated by teachers, not pupils, and a vital message of the Thinking Moves A-Z is to encourage pupils to think independently. Every move is, in effect, the answer to a question. It is good if a pupil gives a thoughtful answer to a teacher's question. It is better still if they stimulate their own thinking by asking themselves more questions.

Synonyms	Alternative Synonyms		Intellectual Virtue
Inquire	Where, when, etc	(Re)search	Inquisitiveness
Investigate	Open/close	Conceptual	
Puzzle	Problem	Inquiry	
Problematize	Awe	Leading	
	Mystery	Rhetorical	
	Empirical		

02

"提问" 步法释义

年幼的学生更擅长提问，从孩童期过渡到青春期后，这种技能似乎在逐渐消失。学生需要通过不断地练习来培养探究的意愿和技能：通过后天练习将提问培养成一种日常习惯，但也要鼓励那些本能的提问。

有时，教师可能会以鼓励（或更好的方式）来回应一些私人研究。除此之外，一个探究性问题也能协助教师鼓励并赋能学生的独立思考。

课堂上的提问通常是由教师主导的，思考步法 A—Z 想要传达的一个重要信息，就是教师需要鼓励学生独立地思考。每个思考步法实际上都提供了一种解答问题的方式。我们希望看到学生能够对教师的问题给出经过深思熟虑的答案，但更令人期待的是，他们能通过更多的自我提问来激发自己的思考。

同义动词	相关词汇		智力美德
问询	时间，地点等	（重新）搜索	求知欲
调查	开放 / 封闭	概念上的	
解决	问题	调查	
问题化	敬畏	领先的	
	神秘事物	修辞的	
	经验的		

03

ACTING PLAN

Question Chain

Partners take turns asking each other questions, aiming to link them all in a chain by ensuring that at least one word from one question appears in the next one. Importantly, the "question" word in one question (e.g. when, how, is, could) should NOT start the next one. This rule ensures a greater variety of questions and chooses the most interesting chain of three to read out.

Four Wonders For

This can be used at any time in a topic or unit but the earlier the better. Ask students to each come up with four "I wonder … ?" statements about the topic. These should be four things they are curious about and to which they don't know the answer.

Ask Me a Question

Turn the tables and get the students to ask you questions about the topic. Don't ask, "does anyone have a question?"—you probably won't get any. Ask, "Could everyone write down three questions to ask me: the one you

03

步法应用

问题链

　　学生两两结伴，轮流向对方提问，目标是将所有问题进行链式连接，确保至少有一个在上个问题中包含的词语要出现在下一个问题中。还有一条重要规则是，上一个问题中的提问词（例如：何时、如何、是、能否）不能用来引发下一个提问，这条规则确保了问题的多样性。最后请选出最有趣的三条问题链分享给大家。

四个好奇点

　　这个活动可以在任意主题或单元的任意一个阶段使用，但放在起步阶段可能更好。让学生每人提出四个关于主题的好奇点，用"我想知道……？"的句式进行陈述。这应该是他们最感兴趣但并不知道答案的四件事。

请问我一个问题

　　反向视角，让学生成为提问者，向教师提出关于当下学习

think will be most helpful to your learning about this topic; the one you think is the most interesting; and one you think will be difficult to answer but worth answering." Write them on whiteboards, pick some to answer and show you don't know everything, enlisting their help. It's a good way of showing vulnerability and curiosity.

Applications

Applications
Give me four wonders for ...

History
Give me four wonders for the Romans (e.g. why was their army so strong?)

Science
Give me four wonders for plants (e.g. can plants survive in space?)

Math
Give me four wonders for numbers (e.g. what's the biggest number?)

Wise Saying

There is more to be learned from the unexpected questions of children than from the discourses of men.

——John Locke

主题的问题。不要用"有人有问题吗?"这样的句子提问,你可能得不到任何回应。可以使用这样的指令:"请每个人写下三个问题:一个你认为对了解这个主题最有帮助的问题;一个你认为最有趣的问题;还有一个你认为很难但值得回答的问题。"将学生的问题收集起来写在白板上,挑一些进行回答,也可以请求学生的帮助,告诉他们你也并不是什么都知道。这是一个很好的方式,适当地示弱,表达你对问题的好奇心。

学科中的应用

应用
请告诉我你关于……的 4 个好奇点

历史
请告诉我你关于"罗马人"的 4 个好奇点(例如:为什么他们的军队如此强大?)

科学
请告诉我你关于"植物"的 4 个好奇点(例如:植物能在太空中生存吗?)

数学
请告诉我你关于"数字"的 4 个好奇点(例如:最大的数字是多少?)

名人语录

从孩子们的意想不到的问题中学到的东西比从成人的交流中学到的更多。

——约翰·洛克

R
RESPOND

回应

鲸　鱼

Relating Animal: Whales

关联理由：为彼此唱歌

Relating Reason: Sing to each other

01

WHAT IS *RESPOND*?

Synonyms	Coaching Questions
Answer	Who has a response to that?
Reply	Is that an answer to the question?

People often say you should live life to the full. Usually, they are encouraging others to take opportunities to go places or pursue some activity. But part of getting the most out of life is responding with interest to what is going on around you. It is good to be an active listener and observer, but it is better still to respond with an answer or at least with an expression of interest.

01

什么是"回应"?

关键同义词	指导问题
回答	谁对此有回应?
答复	这是对问题的回答吗?

人们常说,应该充实地生活。通常,这是在鼓励别人抓住机会去旅行或进行某些活动。但充实生活其实还包括饶有兴趣地回应周围所发生的事情。尝试着做一个积极的倾听者和观察者,显示出对事物的兴趣或给出回应将更为有益。

02

EXPLAIN *RESPOND*

One of the regular complaints of teachers is that some of their pupils are passive learners; they are not interested in "learning" other than what they have to know for tests or exams. At its worst, this is the opposite—even the nadir—of education.

Good teachers manage to stir the interest of their students in their subject, and the best teachers are those who stir interest in the relationship between their subject and the project of humanity. But how can this be done?

There might be as many answers to this question as good teachers, but one simple thing might link them all: showing interest in what interests students. This is not a facile interest in student pastimes—though that could sometimes be the basis for a growing, intellectual relationship. It is rather a matter of modeling a constructive response to any response that students make to the subject matter in hand: asking probing questions or encouraging conceptual connections to be made. In short, it is about dialogical, or philosophical, teaching. (see www.philosophicalteaching.com)

Synonyms	Alternative Synonyms		Intellectual Virtue
React	(Dis) Like	Opinion	Responsiveness
Remark	Made me	Comment	
Feed back	(I) Think	Personal	
Comment	Reaction	Subjective	
	Feeling	View	
	Emotional	Value	

02

"回应"步法释义

教师经常抱怨部分学生学习非常被动；除了考试必须知道的内容之外，他们对"学习"毫无兴趣。最糟糕的是，这与教育的本质是背道而驰的。

好教师能够激发学生对所学学科的兴趣，而最优秀的教师则是那些能激发学生关联学科知识与人类社会兴趣的人。但如何才能做到这一点呢？

好教师有多少，这个问题的答案可能就有多少，但这些答案可能有共通之处：去关心学生感兴趣的事情。虽然有时候那些消遣时间的、肤浅的兴趣也可成为智力关系不断发展的基础，但在此，我们更多指的是所学主题中学生可能感兴趣的内容。不管学生当前的反应如何，教师都可以做出建设性的回应——提出探究性问题或鼓励建立概念联系，来帮助学生继续挖掘。简而言之，这是对话式的或哲学化的教学。（参见 www.philosophicalteaching.com）

同义动词	相关词汇		智力美德
反应	（不）喜欢	观点	响应性
评论	让我	评论	
反馈	（我）认为	个人的	
评论	反应	主观的	
	感觉	视角	
	感性的	价值	

03

ACTING PLAN

Yes, and ... Yes, but

One student in a pair starts with a statement "of the moment" —perhaps an observation, a recollection, or a point of view. The partner must respond, beginning, "Yes, and ..." and finishing with a complementary statement. The first student, in turn, responds with, "yes, and ..." and so on. An alternative or additional format is for the response to begin, "yes, but ..." An interesting difference between the two sorts of responses can emerge when students reflect on their conversations.

Sentence Sharing

One student in a pair gives one word to start a sentence, e.g. "Yesterday" or "sometimes". The partner gives the next word, the first student the third—alternating words but with the aim of finishing the sentence with the 10th word. Ideally, students count the words on their fingers. Sentences can turn out to be fairly plain or quite surreal, and the point is to respond constructively to each other.

All Hands on Deck

Discussions are more thoughtful and go deeper when students respond

03

步法应用

是的，而且……，是的，但是……

两两结伴，学生 A 以关于"当下"的陈述开始——可能是一个观察、回忆或观点。学生 B 必须以"是的，而且……"开头回应，并对观点进行补充陈述。学生 A 再次以"是的，而且……"回应，依此类推。另一种活动玩法是：回应以"是的，但是……"开始。当学生反思自己的对话过程时，他们会发现这两种回应之间的有趣差异。

句子分享

两两结伴，学生 A 用一个词语开始一个句子，如"昨天"或"有时"。学生 B 给出下一个词，学生 A 再给出第三个词语——两人轮流说词，但目标是用 10 个词完成句子。学生最好边说边用手指记数。最终完成的句子可以很简单，也可以相当超现实，活动的重点是互相做出了建设性的回应。

to one another rather than make isolated points directed at the teacher. In a discussion, ask everyone to respond to a point made by indicating if they agree or disagree. Better than thumbs up or down is to have hands on the table for agreement, under the table for disagreement. Some will discover that one hand on the table and one under can stand for "I agree and disagree". Then you can hear the reasons why people agree or disagree with the statement.

Applications

Applications
All hands on deck ...

Math
Is Julie's answer correct?

History
Martin has said Henry VIII was a bad man but a good king. Do we agree?

Art
Usamaah thinks his work doesn't need anything more. What do we think?

Wise Saying

Life is a gift, and I try to respond with grace and courtesy.

———Maya Angelou

全员参与

　　当学生开始相互回应，而不只是分别对教师进行回应时，讨论会变得更有思想，更为深入。在讨论中，可以要求每个人都对某个观点进行回应，通过表示同意或不同意给出自己的立场。学生可以用大拇指朝上或大拇指朝下的手势来表示自己的选择，还有一个可能更好的方式是，把双手放在桌上表示同意，把双手放在桌下表示不同意。有的学生会将一只手放在桌上、一只手放在桌下，这代表"我既同意又不同意"。接着请大家分享自己的理由。

学科中的应用

应用
全员参与

数学
朱莉的答案正确吗?

历史
马丁曾说过亨利八世是个坏人，但是个好国王。大家同意吗?

艺术
乌萨马赫认为他的作品不需要更多的修饰了。大家怎么认为?

名人语录

生命是一份礼物，我努力以优雅和礼貌来回应。

——玛雅·安吉洛

S

SIZE

度量

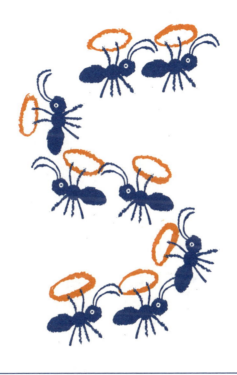

关联动物

蚂　蚁

Relating Animal: Ant

关联理由：测量（对大小有感知）

Relating Reason: Measuring (aware of size)

01

WHAT IS *SIZE*?

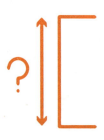

Synonyms	Coaching Questions
Estimate	Could we try to put a figure on how many?
Quantify	Are you saying all, or most, or just some?

There will be many occasions when you need to know, roughly, the number, amount or size of something. To do this you need to quantify. Quantifying may be calculating exactly, or just making a good estimate so you've a good idea of how many of something there are. If you are not careful, you can make big mistakes—not ordering enough food for a party, for example!

01

什么是"度量"?

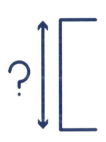

关键同义词	指导问题
估计	我们能试着估计一下具体有多少吗?
量化	你是说全部、大多数,还是只有一些?

在很多时候,你需要了解某物大致的数、量或大小,为此,你需要进行量化。量化可以是精确的计算,或仅仅是做一个尽量准确的估计。如果估量的差异很大的话,可能导致比较严重的问题——例如,没有为派对订购足够的食物!

02

EXPLAIN *SIZE*

"Getting the measure of something" or "sizing up the task" is essential for planning ahead. Estimating how much equipment you will need, or how long a task might take, is a routine part of most practical projects. Reckoning how much the average person might eat, and how much it might cost, informs how much of a party budget goes on food. Computing how long a journey might determine which mode of transport is taken or whether you stop overnight.

There is also a clear value to students in deliberately practicing this move at school. Estimating how long this week's homework might take will impact on their efficiency (as well as what social plans can be made). Reckoning how long to spend on each question in a test can pay dividends. And, of course, the more advanced most studies get, the more students need to appreciate the significance of statistical data.

Synonyms	Alternative Synonyms		Intellectual Virtue
Count	All/No (ne)	Fraction	Sense of proportion
Measure	Some/Most	Proportion	
Reckon	Few/Many	Majority/minority	
Figure	Number	Scale	
	Amount	Degree	
	Frequency	Continuum	

02

"掌握事物的尺度"或"评估任务的大小"是提前规划的重要一环。估计你需要多少设备，或完成一个任务可能需要多长时间，是我们在做大多数项目时要进行的常规步骤。估算每个人的平均食量以及食物的花费，就能决定派对预算中有多少可以用于食物。计算一趟旅程的时长可能决定了你会采用哪种交通方式，或者你是否在旅途中过夜。

对学生来说，在学校里有意识地练习"度量"的步法是很有意义的。估计完成本周的家庭作业需要的时间将影响他们的效率（以及可以制订什么样的社交计划）。计算考试中每个问题上需要花费的时间也是有好处的。当然，随着学习者进入更高的学习阶段，他们可能会认识到估量数据变得更加重要了。

同义动词	相关词汇		智力美德
计数	所有 / 都没有 /	份额	分寸感
测量	一些 / 大多数	比例	
估计	几乎没有 / 很多	多数或少数	
计算	数字	等级	
	总数	程度	
	频率	连续体	

03

ACTING PLAN

Figure it Out

Making reasonable estimates of quantities is a useful skill. This can be as simple as guessing the number of marbles in a jar or estimating the number of chairs in the school. For older students, as a greater challenge, you can create a "Fermi question", named after the educated guesswork the famous physicist used to estimate the chances of intelligent life elsewhere in the universe. For example:

How many Camembert cheese are there in France?

How many Math teachers are there in the UK?

How many goals were scored in the Premier League last year?

All or Nothing

Introduce students to the form of logical (strictly, Aristotelian) statements: "All/Some/No As are/have Bs" and offer a first example: "All trees have leaves". Invite a next statement beginning "Some/No leaves", e.g. "No leaves are made of metal," and then, e.g. "Some metals are used to make coins."

03

步法应用

弄清楚

合理地估计数量是一项很有用的技能。猜测罐子里有多少个弹珠，估计学校里有多少把椅子，有时候这个技能的应用非常简单。而对于更高年级的学生来说，你可以提高应用的难度，比如创建一个"费米问题"[①]，估算宇宙中存在其他智慧生命的概率：

例如：法国有多少卡门贝尔奶酪？

英国有多少数学老师？

去年英超联赛中总共进了多少球？

全有或全无

向学生介绍逻辑陈述的形式（更严格来说，是亚里士多德式陈述）："所有 / 某些 / 没有 A 是 / 有 B，并提供第一个例子："所有树都有叶子"。邀请下一个陈述以"某些 / 没有叶子"开

[①] "费米问题"因意大利裔美国物理学家恩里科·费米而得名，通常指的是一类估算问题，要求在没有精确数据的情况下，通过对复杂问题进行一系列合理的假设和拆解，通过有根据地估算来得出答案。——译者注

Rule 1: the quantifier (All/Some/No) must change each time.

Rule 2: the rest of the sentence must contain a noun, which can then be re-used in the next statement. The game can be played in groups of any size.

A Number to Think of

You can generate curiosity and engagement around any topic by giving a number that is connected to the topic in some way and asking what it's the number of.

Applications

Application
Number to think of

Geography
13 mm (average growth of a stalactite) 13 mm

Chemistry
450 million (tons of fertilizer produced using the Haber process annually)

Math
23 249 425 (number of digits of the largest prime number discovered as of July 2018)

Wise Saying

Measure twice, cut once.

Proverbial advice

始，例如"没有叶子是由金属制成的"，然后例如"某些金属用于制造硬币"。

规则 1：每次都必须改变量词（所有/某些/没有）。

规则 2：句子的其余部分必须包含一个名词，这个名词可以在下一个陈述中重复使用。这个游戏可以在任何规模的团队中进行。

一个值得思考的数字

你可以通过设置一个与当前主题相关的数字，请大家进行猜测。以此来激发大家对该话题的好奇心和参与度。

学科中的应用

应用
值得思考的数字

地理
13 毫米（钟乳石平均生长）

化学
4.5 亿吨（每年使用哈伯法生产的化肥）

数学
23 249 425（2018 年 7 月发现的最大质数的位数）

名人语录

测量两次后再去行动。（三思而后行）
坊间谚语的建议

T

TEST

试验（验证）

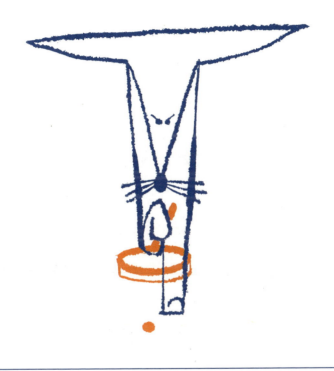

关联动物

狐　狸

Relating Animal: Fox

关联理由：怀疑精神

Relating Reason: Spirit of doubt

01

WHAT IS *TEST* ?

Synonyms	Coaching Questions
Doubt	Is there any reason to doubt this claim?
Check	Let's check the assumptions/facts!

Not all ideas are good ones, and it is often wise to put them to the test—that is, to check whether they are well-based or well-thought though. Just like a spelling test tests how good at spelling you are, to "test" an idea means to see how good it is.

01

什么是"试验（验证）"?

关键同义词	指导问题
怀疑	有理由怀疑这个说法吗?
检核	让我们检查一下这个假设 / 事实!

　　并不是所有的想法都是好想法，选择去进行验证通常是没错的——检查它们是否有充分的依据或是否经过了深思熟虑。就像拼写测试检验你的拼写能力一样，"验证"一个想法意味着看看它究竟是不是一个好想法。

02

EXPLAIN *TEST*

There are different sorts of ideas, and they can be tested in different ways. One main sort is beliefs about the world—what we claim or Maintain to be true. These are usually tested or challenged by calling for evidence or simply by checking the facts, including assumptions, for oneself.

This is a move that might be particularly encouraged in budding scientists, but it is not the prerogative only of scientists. In fact, it is a practice that should be encouraged in every subject and in everyday life. Careful testing remains the bulwark against false gossip, urban myths and, of course, fake news!

Another important sort of idea is a value judgment—not so much a claim about how the world is as about how it ought to be, or what we value and try to promote. When we "test for value", we are not primarily calling for facts but for justifications—of principles, proposals, opinions. We don't assess these for their truth, but on a scale from the strongest possible reason to extremely weak ones.

Synonyms	Alternative Synonyms		Intellectual Virtue
Put in question	Doubtful	Bias	Skepticism
Make sure	Claim	Checklist	
Challenge	Really	Mistake (n)	
Examine	Sure	Testimony	
	Questionable	Reliable	
	Assumption	Confirmation	

02

"试验（验证）"步法释义

想法有不同的类型，我们可以用不同的方式来进行验证。一种主要的类型是关于世界的信念——那些我们声称或坚持认为是真实的东西。验证或挑战这类想法时，我们通常通过要求提供证据或核验事实来进行，当然也包括自验假设。

这个步法可能是崭露头角的科学家特别推崇的，但它不仅是科学家的特权。事实上，应该在每个学科和日常生活中鼓励这种实践。仔细的测试和谨慎的验证仍然是抵御虚假流言、都市神话，当然还有假新闻的坚固屏障！

另一类重要的想法类型是价值判断——和上一类关于"世界是什么样子"的主张不同，这一类主张更关心的是"世界应该是什么样子"，或者是关于我们重视什么、试图提倡什么。"检验价值"时，我们主要做的不是核验事实，而是要求辩理——包括核验原则、建议和意见的正当性。我们不评估它们的真实性，而是从其理由的有力程度来进行评估。

同义动词	相关词汇		智力美德
提出问题	生疑的	偏见	怀疑主义
确保	要求	检查表	
挑战	事实上	错误（被误解的）	
检查	应当如此的	证词	
	有问题的	可靠的	
	假设	确认	

03

ACTING PLAN

How Would You Test?

Come up with a range of statements that can be tested for truth. They can be both fantastical and factual—the important thing is they are beyond everyday experience. In pairs, get students to suggest how they would carry on the test, step by step. Examples: "How would you test ... if there were an edge to our universe? ... if unicorns once existed? ... if aliens had abducted you last night and put you in a classroom simulation?"

Lie Detector

A development of True of False (see Maintain). At the beginning of any new topic, give students two statements, one true, one false, or one statement that can be either true or false. Ask them to pair up and work out how they would test which is true which is false. What information would they need? What questions would they ask? You might even set them the task of finding out the answers.

Verify/Falsify

Put a range of true and false statements on the board. Ask students to

03

步法应用

你会如何试验？

提出一系列可以被验证的陈述。它们既可以是幻想的，也可以是关于客观存在的——重要的是，它们是超出日常生活经验的。学生两人一组，让他们提出如何一步一步进行这个测试。例如："你会如何试验我们的宇宙是否有边界？你会如何检验独角兽是否曾经存在？外星人昨晚是否绑架并将你放在一个模拟教室中，你会如何检验？"

测谎仪

该活动是"保持"一节中的活动"真或假"的进阶版。在开始任意新主题之前，向学生出示两个陈述，一个为真，一个为假，或者一个可以为真也可以为假的陈述。让学生两两结伴，判断哪个是真，哪个是假。看看他们是如何进行判断的？需要哪些信息？会问什么问题？你甚至可以不出示答案，请他们自己去寻找。

work through then, using their knowledge to either Verify (prove it to be true) or Falsify (prove it to be false), showing their working in each case. If they cannot verify or falsify, they should explain why and outline what could assist them, or how they might assess or size the degree of probability. They could also argue that the statement is a value judgment and therefore neither verifiable nor falsifiable.

Applications

Application
How would you test ...

Geography
How would you test if Brazil's economy is thriving?

Math
How would you test which of these equations is correct?

Music
How would you test if the moods music creates are natural or learned

Wise Saying

Good tests kill flawed theories; we remain alive to guess again.

——Karl Popper

证实／证伪

在黑板上写出一些真实和虚假的陈述。让学生通过使用自己的知识来证实（证明其为真）或证伪（证明其为假），并展示他们在每条陈述下的思考过程。如果不能证实或证伪，应该解释原因并概述可能带来帮助的信息，或者他们可以如何进行评估或衡量可能性。他们也可以认为该陈述是一个价值判断，因此既不能证实也不能证伪。

学科中的应用

应用

你会如何试验……

地理

你会如何试验巴西经济是否正在蓬勃发展？

数学

你会如何检验下列哪个等式是正确的？

音乐

你会如何试验音乐创造的情绪是自然的还是习得的？

名人语录

好的试验会扼杀有缺陷的理论；我们可以一直验证下去。

——卡尔·波普尔

U

USE

使用（工具）

关联动物

乌 鸦

Relating Animal: Crow

关联理由：智慧且注重实用的（乌鸦喝水故事利用工具）

Relating Reason: Intelligent and Practical

(The Crow and the Pitcher)

01

WHAT IS *USE*?

Synonyms	Coaching Questions
Try out	Who can see a use for this?
Apply	Let's put this into practice!

If an idea looks like a good one, then it makes sense to put it into practice, otherwise you might as well not have had it. This means looking out for the opportunity to use it—and taking that opportunity!

01

什么是"使用（工具）"？

关键同义词	指导问题
试用	谁知道这有什么用？
应用	让我们将它付诸实践！

如果一个想法看上去不错，那就将它付诸实践吧，不实践，则无以立。寻找并抓住尝试的机会！

02

EXPLAIN *USE*

Whatever we do with our minds can be put to good use. It might be knowing something that helps us navigate our way through the world, or a skill that can be applied in various contexts once mastered. Or it might be an idea or principle arrived at from discussion that can then be put into action, such as following through on a discussion about freedom of speech by supporting a charity for imprisoned journalists.

Of course, learning needs to be put to use inside the classroom too. Knowledge and skills can fade and rust if they're not regularly applied. We should ensure that students devote enough of their lesson time to demonstrating, transferring and using what they've taken on. The question we, and they, might continually be asking is: "How can we apply this understanding?"

Synonyms	Alternative Synonyms		Intellectual Virtue
Try out	Practical	Function	Pragmatism
Put to use	Experiment	Utility	
Experiment	(In) practice	Purpose	
Implement	(In) action	Implementation	
	Effective	Transfer	
	Application	Imitation	

02

"使用（工具）"步法释义

　　不论想做什么，开动脑筋能帮你实现良好的"使用"实践。可能是能帮助我们遨游世界的航海知识，可能是一项可迁移的技能。或者是从讨论中得出的一个可以付诸行动的想法或原则，比如在讨论言论自由后，通过支持一家帮助被监禁记者的慈善机构来践行这一讨论。

　　当然，课堂中的学习也需要"使用"。如果不经常运用，知识和技能可能会褪色和生锈。我们应确保学生把足够的课堂时间用于展示、迁移和使用他们所学的东西。一个可能被不断提出的问题是："我们如何应用这种理解？"

同义动词	相关词汇		智力美德
考核／试用	实用的	功能	实用主义
投入使用	实验	实用	
实验	（在）实践	目的	
实施	（在）行动	执行	
	有效的	迁移	
	应用	模拟	

03

ACTING PLAN

Message in Your Pocket

Each student writes down on a piece of paper a message to themselves which they think would be useful to remember during the week. It could be something about how to act, how to think, how to feel. They then fold the paper up small, and keep it close to them, e.g. in a back pocket, to remind them of their intention. At the end of the week, ask them to think about whether and when they used their message in a Pocket. For best results, do this exercise yourself with something the week before, and share the result with them beforehand.

Who and How

At the end of any period of study (could be a lesson or a whole unit), ask students to reflect on the skills and knowledge they have developed. Who in the outside world would use this? How would they use it? This not only helps them think about applications and uses, but it also helps them appreciate the relevance of what they are learning (and might provide some career advice by stealth)!

So, What?

At the end of a discussion (about a problem facing us or about the rights

03

口袋里的纸条

请每个学生在一张纸上写下一条他们认为在一周内会有用的信息。它可以是关于如何行动，如何思考或者如何感受的信息。然后，把纸折叠起来，放在一个身边随时能想起来的地方，比如裤子后袋里。在一周结束的时候，让学生思考自己是否以及何时使用了口袋里的信息。为了达到最好的效果，你可以提前一周自己先进行这个练习，并和学生分享结果。

谁以及如何

在任何一个学习阶段（可以是一节课或一个单元）结束时，要求学生反思他们已经掌握的技能和知识。现实生活中，可能还有谁会用到这些技能和知识？他们会如何使用？这个活动不仅可以帮助学生思考知识的应用，还能帮助他们了解所学内容的相关性（还可能为他们提供一些职业建议）！

所以，接下来呢？

在讨论结束时（关于我们面临的问题或关于某事的对与

and wrongs of something), it would seem a waste of time and energy if we then go back to business as usual. End a discussion by asking the class "So, what?", and let them suggest applications of principles and conclusions they've arrived at.

Choose a Move

Help learners practice thinking about other Thinking Moves in a fun and fantastical way. Present them with a scenario, such as below, asking what Thinking Move they would need to make ...

... if there was a power shortage so that there was only electricity for three hours each day?
... if they discovered they had shrunk to the size of a Lego person?
... if they won a million pounds but had to spend it all within one week?

Applications

Applications
So, what ...

Geography
So, what will you do to reduce plastic pollution?

PSHE
So, what will you do to help new students settle in?

Art
So, what will you do to incorporate impressionism into you work?

Wise Saying

The great end of life is not knowledge but action.

——Thomas Henry Huxley

错），如果我们仍然回到往常的样态中，似乎讨论就是一种对时间和精力的浪费。可以在讨论结束时，询问全班同学："所以，接下来我们该怎么做？"请他们对已经得出的原则和结论提出实施建议。

选择一个步法

帮助学习者以一种有趣和新奇的方式练习使用其他的思考步法。向他们呈现一个场景，询问他们会使用哪个（哪些）步法来进行思考。例如：

如果出现电力短缺，每天只有三个小时的电力供应？

如果发现自己变得和乐高小人一样大？

如果你赢得了一百万英镑，但必须在一周内全部花掉？

学科中的应用

应用
所以，接下来……

地理
所以，接下来你们会减少塑料污染吗？

个人、社会和健康教育
所以，接下来你会做什么来帮助新生适应？

艺术
所以，接下来你会做什么来增加作品中印象派的色彩？

名人语录

生命的伟大目标不是知识，而是行动。

——托马斯·亨利·赫胥黎

VARY

变通

蝴　蝶

Relating Animal: Butterfly

关联理由：由毛毛虫变化而成

Relating Reason: Changing from Caterpillar

01

WHAT IS *VARY* ?

Synonyms	Coaching Questions
Change	Is there a better way of doing this?
Alter	Does anyone have an alternative idea/perspective?

Sometimes we all just get stuck. We know things are not working out, but we don't know what we should do about it. A very good next move, then, is quite simply to try something different!

01

什么是"变通"?

关键同义词	指导问题
改变	有更好的方法来做这件事吗?
更改	有人有另一种想法 / 观点吗?

　　有时候我们会陷入困境。我们虽然知道事情没有解决,但却不知道该怎么办。那么,接下来一个很好的做法,非常简单,就是去尝试一些不同的东西。

02

EXPLAIN *VARY*

VARY is a prime creative Thinking Move. It can be made at the everyday level of striving to improve whatever one is doing—including what one is saying or writing. "what went well, even better if" is a good strategy of this. Or it can be made at the more entrepreneurial level of trying to come up with radically new ideas.

A great example of this in pedagogy is Pie Corbett's Imitate, Innovate, Invent approach to writing: first becoming familiar with a text, and then varying elements of it, before more independent writing. It's an important contribution to undoing the bad press that "copying" traditionally has in schools.

Varying can involve a trial-and-error approach, i.e. trying different things and seeing how well they work—"playing around" with an idea. Or it can involve rethinking an approach—"thinking laterally"—most famously done by the likes of Darwin and Dyson. Either way, varying discourages learners from settling with their first answer or idea.

02

"变通"步法释义

变通是一个代表创新式思维的思考动作。它可以提升日常生活中各种事情的完成质量——包括口头表达或书面写作。"哪些地方进展得较好,哪些地方调整一下会更好?"是一个很好的策略。而从更具创新性的层面来看,此步法也代表着尝试提出全新的想法。

派·科贝特创设的写作三部曲"模仿、创新、发明"就是教学法一个很好的例子:在进行更多独立写作之前,先熟悉一篇文本,然后变化它的元素。这对改变学校传统上对"抄袭"的不良看法做出了重要贡献。

变通,可以是内容的变通,你可以进行多次试验来尝试不同的内容,看看它们的效果如何——但始终围绕着一个想法,就像与它玩耍一样。也可以是方法的变通,重新思考并选择另一种方法,像达尔文和戴森一样,尝试进行"横向思考"。无论哪种方式,变通都鼓励学习者不要满足于他们的第一个答案或想法。

Synonyms	Alternative Synonyms		Intellectual Virtue
Play with	Different	Version	Adaptability
Adapt	(An) other	Extra	
Modify	Way	Substitution	
Diversify	Perspective	Modification	
	Instead	Adaptation	
	Alternative	Trial and error	

同义动词	相关词汇		智力美德
与……玩耍	不同的	版本	适应性
适应	其他	额外的	
修改	方式	替代物	
（使）多样化	视角	修改	
	代替	改良	
	供替代的选择	试错法	

03

ACTING PLAN

New Choice!

A paired storytelling game. One player starts telling a story. At any time, their partner can call, "New Choice" and the last thing said or done must be changed.

A: Once upon a time there was a boy ...

B: New Choice!

A: Once upon a time there was a girl ...

B: New Choice!

A: Once upon a time there was a hippo. He had lots of hippo friends who liked wallowing in the mud ...

B: New Choice!

Socrates' Sieve

1. Start with a proverb or popular opinion, such as, "It's wrong to hurt people."
2. Find an exception where is doesn't work ("A nurse giving an injection").
3. Restate the principle with qualifiers that allow for the exception ("It's wrong to hurt people unless you are doing it for their own good").

03

新选择！

两两结伴讲故事的游戏。由学生 A 先开始，在讲述过程中，学生 B 可以通过喊出"新选择"来中止 A 的讲述。听到这个口令时，A 就必须改变刚才故事中出现的最后一句话。

A：从前有个男孩……

B：新选择！

A：从前有个女孩……

B：新选择！

A：从前有一头河马。它有很多喜欢在泥里打滚的河马朋友……

B：新选择！

苏格拉底的筛子

1. 从一句谚语或普遍观点开始，例如："伤害他人是错误的。"

2. 找到一个不适用的例外（"护士打针"）。

3. 重新陈述观点，加上使这个反例成立的限定条件（"伤害

4. See if there is an exception to the restated principle ("A parent spanking a child").

5. Repeat the process until you have a principle that covers all situations.

Give Us Another!

This can be used after asking any question that has several possible, valid answers and is a playful way to stretch confident speakers. Ask a question and if such a student answers, fire back, "Give us another!" and keep going until they run out of alternative answers. The answers must be correct (they can't just give wrong ones) and if there are still right answers that remain unsaid, invite others to "Give us another!"

Applications

Application
Give us another ...

Geography
Physical process

Technology
Way to cut wood

History
Consequence of the slave trade

Wise Sayings

When we are no longer able to change a situation, we are challenged to change ourselves.

——Vikor E. Frankl

Change your life today. Don't gamble on the future, act now, without delay.

——Simone de Beauvoir

他人是错误的，除非你是为了他们的利益"）。

4. 查看是否有不符合重新陈述后的观点的例外（"父母打孩子的屁股"）。

5. 重复这个过程，直到产生涵盖所有情况的陈述。

再来一个！

当你抛出一个有多种可能或多个有效答案的问题时，这将是一个很有趣的活动方式，活动能大大鼓励发言者的自信。在教师给出问题后，如果有学生给出答案，你可以迅速回应："再来一个答案！"并持续进行，直到他们没有其他答案为止。学生给出的答案必须是正确的（不能只是给出答案而不考虑答案的质量），如果还有未说出的正确答案，可以继续邀请其他人作答。

学科中的应用

应用
再来一个！

地理
物理作用

技术
伐木方法

历史
奴隶贸易的后果

名人语录

当我们无法改变现状时，我们的挑战是需要改变自己。

——维克多·弗兰克尔

从今天开始改变你的生活。不要把宝押在未来，现在就行动，不要拖延。

——西蒙娜·德·波伏娃

权衡（选择）

关联动物

猫头鹰

Relating Animal: Owl

关联理由：**表现出良好的判断力和智慧**

Relating Reason: Showing good judgment / wisdom

01

WHAT IS *WEIGH UP*

Synonyms	Coaching Questions
Decide	Let's weigh up the pros and cons for this.
Judge	So, what has been decided?

This is what you do, or least should do, whenever you make a decision—weigh up the reasons for and against something. Some choices are obvious, but others can be close calls that take a lot of careful thinking about pros and cons.

01

什么是"权衡（选择）"？

关键同义词	指导问题
决定	让我们来权衡一下利弊吧。
判断	那么，已经决定了什么？

这就是你在做决定时所做的，或至少是应该做的——权衡赞成和反对某事的理由。有些选择是显而易见的，但有些则难以选择，需要仔细考虑利弊。

02

EXPLAIN *WEIGH UP*

It's no coincidence that the most common symbol for judging is old-fashioned scales. The scales vividly represent how we weigh up the evidence on each side of a dispute, or the pros and cons of a decision, to reach a balanced judgment. That can be through objective measurement or complex, often subjective, criteria. Of course, deciding what criteria to use is itself a matter of judgment, as is deciding which criteria are the most important.

To judge is to be aware that not all answers are obvious. Some questions are contestable, and so their answers won't (and shouldn't) be provided by the teacher. Most formal assessments or exams require independent conclusions to be drawn, but judging is important beyond the exam hall. Wise decisions are needed in personal and professional lives about what is true, what is right, and what to do. These involve careful thought and consideration of our own biases too.

Synonyms	Alternative Synonyms		Intellectual Virtue
Choose	Good/Bad	Dilemma	Judiciousness
Assess	Best/Worst	Balance	
Evaluate	Right/Wrong	Impartial	
Deliberate	(Would you)Rather	Considerations	
	Decision	Criteria	
	Pros/Cons	Verdict	

02

"权衡（选择）"步法释义

　　判断最常用的表示符号形似一个老式天平，这并非巧合。天平生动地表现了争论双方的证据如何被权衡，或某个决定的利弊如何被比较，并最终呈现平衡（做出判断）的过程。我们可以通过进行客观的测量或使用复杂的（通常是主观的）标准来实现权衡。当然，决定使用什么标准本身就是在做出判断，因为你需要在诸多标准中确定哪些是最重要的。

　　判断意味着我们需要意识到，并不是所有问题的选择都是显而易见的，有些问题是有争议的，所以教师不会（也不应该）直接提供答案。大多数正式评估或考试会要求你进行判断，得出独立的、正确的答案，但判断在考场之外同样重要。在个人生活和学术/职业生涯中，我们还有更多需要权衡的问题，什么是真实的，什么是正确的，什么是要去做的，都需要我们思考后才能做出明智的决定。权衡包括仔细思考和考虑我们自己的偏见。

同义动词	相关词汇		智力美德
选择	好/坏	两难境地	明智
评定	最好/最坏	平衡	
评估	对/错	公正的	
深思熟虑	（你）宁愿	注意事项	
	决定	标准	
	优点/缺点	判决	

03

ACTING PLAN

Good Idea, Bad Idea

Divide your space in half with a skipping rope. Ask a contestable question, e.g. "Zoos for animals—good idea or bad idea?" Ask everyone to move to the "good idea" side. Get students to pair and share reasons to support the idea. It's like a shoal of fish—safety in numbers! Don't hear back from the pairs or you'll steal the thunder of the final step.

Then move everyone to the "bad idea" side, and do the same with the reasons against. Finally ask them the question again, and let them show what they think now by standing on the side of their choice. Now is the time for some to share their reasons with the whole group, because they have had a chance to safely rehearse what they might say. The same activity, with a suitable question, helps teenagers working on argumentative writing, or you can use it to liven up a longer philosophical enquiry.

Pro-Con Processor

Gather reasons for and against something. Then focus on the "pros", and ask what the strongest reason is on that side of the argument and why.

03

好主意，坏主意

用一根绳子将空间分成两半。提出一个有争议的问题，例如："动物园对动物来说是好主意还是坏主意？"请所有人先移动到"好主意"一边，让学生两两结伴，分享支持"动物园对动物来说是好主意"的理由。这就像是一个鱼群——人多力量大！此时先不过渡到大组分享。

然后让所有人都移动到"坏主意"一边，两两结伴，分享支持"动物园对动物来说是坏主意"的理由。这样，学生就在比较安全的氛围中进行了不同视角的思考，并尝试着进行了语言组织。最后再次提出这个问题："动物园对动物来说是好主意还是坏主意？"请学生进行最终的选择，站在自己更倾向的立场旁边，然后进入大圈分享。教师可以选择更多不同的问题来开展这个活动，该活动有助于青少年进行论证性写作练习，或者用来开展一个更长时间的哲学探究。

正反理由处理器

收集支持和反对某事的理由。先一起关注"正面"（支持该

Continue ranking reasons from strongest to weakest, voting if necessary. Repeat with the "cons". The point of doing each side separately is that students often struggle to separate consideration of the strength of a reason from whether or not they agree with the conclusion it supports.

Find Your MoJO

As indicated in thinking BACK, deliberate reflection is a bedrock of good learning, even at the level of memorizing facts. Deep learning or understanding, though, is promoted by doing something with the facts you can remember. So, plan or be ready to seize Moments of Judgment or Opinion, asking students to make a decision based on the facts that they can remember (or will, when they start thinking about the decision).

Applications

Applications
Pro-con Processor

Geography
Settling of a settlement

History
Was revolution in Russia inevitable?

Wise Sayings

A wise person apportions their belief with the evidence.

——David Hume

Every man must decide whether he will walk in the light of creative altruism or in the darkness of destructive selfishness.

——Martin Luther King, Jr.

事的）理由，思考在这些理由之中，最有力的是什么，以及为什么。从最有力到最弱进行排序，必要时可以进行投票。再关注"反面"（不支持该事的）理由，重复以上做法。从正方和反方的角度分别去讨论的目的在于，帮助学生更好地分辨理由的说服力与个人选择的偏好。

发现判断 / 观点时刻

正如在"回顾"这个思考步法中所指出的，反思是达成良好学习效果的基础，哪怕仅仅是出于巩固记忆的目的。然而，想达到深入学习或理解，我们还需要对能记忆的事实进行进一步的实践。因此，计划或准备好抓住每一个"判断或观点出现的时刻"，鼓励学生根据自己记忆的事实做出决定（或是他们将记住事实，当他们开始考虑做出决定时）。

学科中的应用

应用
正反理由处理器

地理
对定居点的选址

历史
俄国革命不可避免吗？

名人语录

聪明人会根据证据来分配自己的信仰。

——大卫·休谟

每个人都必须决定，自己是在创造性的、利他的光明中行走，还是毁灭性的、自私的黑暗中行走。

——马丁·路德·金

X

eXEMPLIFY

举例

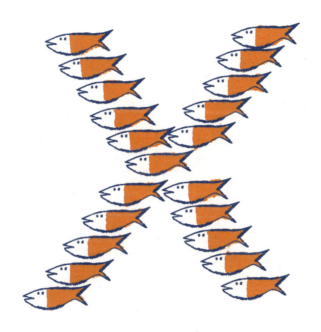

沙丁鱼

Relating Animal: Sardine

关联理由：许多沙丁鱼会聚集在一起

Relating Reason: Many examples

01

WHAT IS *eXEMPLIFY*?

Synonyms	Coaching Questions
Give Example	Can you give an example of that?
Illustrate	Does anyone have a counter-example?

The more complex an idea, the more helpful it can be to give a specific example. It's easier to understand stars by thinking about our own sun. Examples can also be used to strengthen and oppose arguments. People themselves can be thought of as examples. Rosa Parks is often held up as giving an example of courage.

01

什么是"举例"？

关键同义词	指导问题
举例说明	你能举个例子吗？
图示	有谁能提供一个反例吗？

　　我们可以通过举例来理解复杂的想法，想法越复杂，举例就会越有帮助。比如，以自己相对熟悉的"太阳"为例证，我们会更容易理解"恒星"是什么。举例也可以用于加强和反对某个论点。人本身也可以被用来作为例子，例如，罗莎·帕克斯就经常被视为勇气的典范。

02

EXPLAIN *eXEMPLIFY*

Education often proceeds by getting students to understand general principles via memorable examples. How often do you EXPLAIN something in the abstract to someone, to be met with a blank expression, before you say "for example ..." in the hope the penny drops? Examples tie general ideas into specific experience. If we want learners to manage their learning, we should encourage them to think of their own examples.

Examples are crucial not only for understanding but for argument too. In discussions, an example often brings the question to life. It can sometimes be persuasive, but it can equally ignite a lively exchange—especially when it features people or is viewed in different ways. One example often leads to another, perhaps even a counter-example, and that opens up the opportunity for other Thinking Moves such as to CONNECT or DIVIDE the scenarios.

Synonyms	Alternative Synonyms		Intellectual Virtue
Say (for instance)	Example	Case study	Groundedness
Give an instance	Real life	Typical	
Cite	Event	Sample	
Instantiate	Experience	Exemplar	
	Instance	Specimen	
	Scenario	Citation	

02

"举例" 步法释义

　　教育中我们常常会用一些令人印象深刻的例子来帮助学生理解一般性原理。你是否经历过这样的场景：向某人解释一件抽象的事情，但在你说出"比如说……"之前，对方往往一脸茫然。举例将通用原理与具体经验联系起来。如果希望学生能自主学习，应该鼓励他们尝试着自己举例。

　　例子不仅能够帮助理解，对论证也有重要价值。在讨论中，一个例子往往能使问题变得生动起来。有时，举例能增强说服力；有时，举例还可以提高讨论的活跃度——尤其是当它涉及人物，或者出现不同的视角时。一个例子往往会引出另一个例子，甚至可能是一个反例，这就为其他思考步法的使用提供了机会，比如我们需要为问题关联或区分不同的情境。

同义动词	相关词汇		智力美德
说（例如）	例子	案例研究	有根据性
给一个实例	现实生活	典型的	
引用	事件	样本	
实例化	经验	范例／典型	
	实例	标本	
	情境	引述	

03

ACTING PLAN

Example Chain

This paired game makes use of the way the same thing can be an example of two or more general kinds. A asks the questions, B answers:

A: Think of a thing.

B: A seagull.

A: What kind of thing is that?

B: A bird.

A: What's another example of a bird?

B: An eagle.

A: What kind of thing is that? (you can't say "bird")

B: A predator.

A: What's another example of a predator?

B: A tiger.

A: What kind of thing is that?

They can swap roles and share some of the chains of examples and kinds created.

Yes, No, Maybe

For any concept under discussion, ask for something they think is an

03

步法应用

例子链

这个活动告诉我们，同一事物可以作为两个或更多个类型的例子而存在。两两结伴，学生 A 问问题，学生 B 来回答：

A：请想一样东西。

B：海鸥。

A：那是什么类型的东西？

B：鸟。

A：还有什么可以作为"鸟"的例子？

B：鹰。

A：那是什么类型的东西？（你不能再说"鸟"）

B：捕食者。

A：还有什么可以作为"捕食者"的例子？

B：老虎。

A：那是什么类型的东西？（不能再说"捕食者"）

学生可以互换角色并分享他们创造的例子链和类型链。

符合，不符合，不确定

这个活动适用于讨论中出现的任何概念，请学生举出符合这个概念的例子、不符合的例子，以及不确定的是否符合的例

example of it, something they think is not, and something which is a borderline case they're not sure of. It is these last gray examples that provoke the most disagreement and richest discussion.

Needs—Yes, water; no, Xbox; maybe, electricity
Sports—Yes, football; no, baking; maybe, dancing

Best Example

Some things can be held up as paradigms or exemplars of a class or concept. For example, dogs might be regarded as the best example of pets, or London of multicultural cities. Not everyone will agree on a choice, and an interesting discussion could arise as to what criteria are or should be used. Perhaps most interesting could be discussions about exemplary people (leaders, sportspeople, teachers, students).

Applications

Applications
Give us a Yes, No, Maybe for ...

Dance
... a performance

ICT
... a robot

Art
... an artwork

Wise Saying

The road to learning by precept is long, but by example short and effective.

——Seneca

子。往往是那些不确定的、模棱两可的例子能够引发最大的分歧和最丰富的讨论。

概念	符合	不符合	不确定
需求	水	Xbox	电力
运动	足球	烘焙	跳舞

最好的例子

有些事物可以被视为某一类别或概念的典型范例。例如，狗是"宠物"的典型代表，伦敦可能被视为"多元文化城市"的绝佳范例。但关于这个"最好的例子"，并不是所有人都会达成一致意见。不同的评判标准会带来不同的想法，有趣的讨论就随之产生了。试一试给出杰出人物（领导者、运动员、教师、学生）的最佳范例，这可能非常有趣。

学科中的应用

应用
给我们一个符合……的例子，一个不符合……的例子和一个不确定是否符合……的例子

舞蹈
表演

信息通信技术
机器人

艺术
艺术作品

名人语录

通过规则学习的道路漫长，但通过榜样学习短暂而有效。

——塞涅卡

Y

YIELD

让步

短尾矮袋鼠
Relating Animal: Quokka

关联理由：世界上最快乐 / 最友好的动物
Relating Reason: The happiest/friendliest animal in the world

01

WHAT IS *YIELD*?

Synonyms	Coaching Questions
Accept	Has anyone changed their mind?
Concede	What part of their argument can you accept?

You yield when you give way to someone else by, at least, stopping to let them speak. But you yield all the more when you adjust your ideas to theirs. You may even go so far as to change your mind—which you should usually regard as a good thing to do.

01

什么是"让步"？

关键同义词	指导问题
接受	有人改变主意了吗？
退让	你能接受他们论点中的哪一部分？

当你让步，或者至少是暂停表达并认真倾听时，你就是在给予他人表达想法的机会。但更大程度上的让步通常是指根据他人的想法去调整自己的想法，有时候，你做的甚至不仅仅是调整，而是完全改变了自己的想法。这种想法上的转变应该被认可为一件好事。

02

EXPLAIN *YIELD*

The decision to stop holding one belief or position and accept another is clearly a significant move of thought and, often, a collaborative one. It can also be desirable, especially when there is good evidence or argument to warrant a change of mind.

The pity is that humans do not always find it easy to accept when they need to change their minds. They become more attached to "their" ideas than to seeking the truth, although perhaps they just need to save face.

Cultivating a flexibility to see merit in others' ideas, reconcile and reconstruct is a good balance to the habit of "always sticking to what you believe". In teaching activities, a teacher might point to the drift of an argument and ask an individual, gently, if they would be willing to yield some ground. They could also use the language of compromise, pointing out the need sometimes to see a situation from another point of view and be willing to negotiate.

Synonyms	Alternative Synonyms		Intellectual Virtue
Come around to	Having heard	Adjustment	Flexibility
Admit	On second thoughts	Concession	
Accommodate	Maybe	Change of mind	
Compromise	Negotiation	Synthesis	
	Compromise	Self-correction	
	Deal	Reconciliation	

02

"让步"步法释义

不再坚持对某个信念或立场的坚持，转而决定接受其他的信念或立场，这显然是思想的重大转变，而且也往往是一种协作性的转变。当有充分的证据或论证支持时，这种转变是可取的。

遗憾的是，人们并不总是那么容易接受这些需要改变想法的时刻。他们对"自己的"想法的依恋超过了对真理的追求，尽管也许只是出于保全面子的需要。

灵活看待他人观点中的长处，培养协调和重建的能力，这对固执己见的习惯来说，是一种良好的平衡。在教学活动中，教师可能会指出某个论证的大致含义，然后温和地询问某人是否愿意做出让步；也可以使用折中的语言，指出我们有时需要换个角度看问题并保持一种愿意协商的态度。

同义动词	相关词汇		智力美德
回心转意	有听过	调整	灵活性
承认	一转念	让步	
适应	也许	改变主意	
妥协	谈判	合题	
	妥协	自我修正	
	交易	和解	

03

ACTING PLAN

I Used to Think ... Now I Think ...

Ask students in pairs to share something they used to believe, but which they no longer believe. What was it that changed their minds? Gather some of the reasons for the changing beliefs and see what they have in common.

All in All

Ask students individually to list some generalizations—things that they believe are generally true, but which might have some exceptions—with space beneath each one. Next, ask them to think up possible exceptions and add them to the space below. The same activity can be done with suggestions for new laws, and when they would accept them being broken.

Common Ground

This move can be used during or after any debate between two sides. Get each side to identify Common Ground with the other. Good questions to invite this are:

03

步法应用

我曾经认为……现在我认为……

两两结伴，相互分享他们曾经认同但现在不再认同的事情。是什么改变了他们的想法？收集一些让信念发生改变的原因，看看它们有什么共同点。

总而言之

要求学生独立思考，列出一些概括性陈述——指他们认为通常情况下是正确的、但也可能存在例外情况的观点，——在每一个陈述下方留出一些空白区域。接下来，请他们想一想可能出现的例外情况，并写在下方空白处。该活动也可以用于需要对新法规提出建议，以及在何种情况下人们会认可新法规被违背。

共同点

这个步法可以在两方辩论中或辩论结束时使用。让双方试着去找到一个共同点。这里有一些可能有帮助的问题：

What do you agree on?

Where do you accept the other side is right?

What do you concede?

These questions help students realize that one can concede a point without conceding an argument. In each one, the language of the question assumes there is an answer. They can still say, "Nothing! They're wrong about everything!" but the door is open to yielding.

Applications

Applications
Common Ground ...

History
The defence team for Charles I might admit that he has made several mistakes.

Geography
Those arguing for drastic measures to prevent climate disaster might concede that some people might have to give up more than others.

Wise Saying

When the facts change, I change my mind. What do you do, sir?

— Originally attributed to John Maynard Keynes, but we have to concede its origin is unknown!

你们都同意的是什么？

对方的哪一部分观点你认为是对的？

你不得不承认的是什么？

这些问题帮助学生认识到，我们可以在保持自己立场的情况下承认一个观点。这些问题所使用的语言预设了双方的共同点是存在的。即使最终双方仍然坚持自己的立场，说："没有！他们说的都不对！"但是，通过思考这些问题，尝试寻找共同点时，我们已经打开了让步的大门。

学科中的应用

应用
共同点

历史
查尔斯一世的支持者也可能会承认他犯过几个错误。

地理
主张为预防气候灾难采取严厉措施的人也可能会承认，有些人可能不得不比其他人放弃更多的东西。

名人语录

当事实发生变化时，我也会改变我的想法。你会怎么做呢，先生？

——最初我们认为引用自约翰·梅纳德·凯恩斯，但后来我们不得不承认它的起源是未知的！

Z

ZOOM

变焦（聚焦细节 /
关注整体）

鹰

Relating Animal: Eagle

关联理由：俯瞰全局并能根据需求适时聚焦

Relating Reason: Looking over and zooming in

01

WHAT IS *ZOOM*?

Synonyms	Coaching Questions
Focus on	In: What should we focus on now?
Survey	Out: Let's step back and look at the big picture

When directors make films, they use different camera shots for different purposes. They might zoom in to show the sweat on the brow of a hero at a tense moment or zoom out to show the full scale of an army ready for battle. In thinking, zooming in on details and zooming out to get the bigger picture can both be important.

01

什么是"变焦"?

关键同义词	指导问题
聚焦	放大（聚焦细节）：我们现在应该专注于什么？
审视	缩小（关注整体）：让我们退后一步，看看整体情况

　　导演在拍摄电影时，会根据不同的目的使用不同的镜头。有时候可能会拉近镜头，凸显紧张时刻英雄额头上的汗珠，或者拉远镜头，展示一支整装待发、准备战斗的军队全貌。在思考时，放大以聚焦细节，缩小以关注整体，二者都很重要。

02

EXPLAIN *ZOOM*

Zooming in means giving closer attention to the small details. Zooming out shows things in proportion or perspective. The journal of a single soldier or a map of the front can both be illuminating in understanding a conflict. You can zoom in or out in time as well: what makes news today is rarely as important as trends that develop over years and decades.

Choosing the best level at which to investigate something is part of being a good inquirer. You might need to focus on a particular word to understand the question better, but you should also step back occasionally to Weigh up how well your inquiry is going.

In Science, seeing how different levels fit together is especially important to understanding complex systems, e.g. seeing how processes within individual cells connect to bodily movements. Stepping back can also help you spot patterns, which may be significant in History, or Art, as well as in Science.

Synonyms	Alternative Synonyms		Intellectual Virtue
Inspect	Focus	Big picture	Concentration
Scrutinize	Detail	Overview	Comprehensiveness
Review	Particular	General	
Take stock	Tiny	System	
	Aspect	Pattern	
	Perspective	Gestalt	

02

"变焦"步法释义

放大意味着更细致地聚焦细节。缩小则是按比例或以透视的方式来整体性地展示事物。一位士兵的日记或一张前线地图都能为理解战争冲突提供启示。你也可以在时间轴上进行放大或缩小：一则在今天成为重大新闻的事，对数十年的发展大势而言，就可能并没有那么重要。

想成为一名好的探究者，很重要的部分就是得学会选择调查研究的层级及范围。为了获得对所探究问题更好的理解，有时你可能需要专注于某个特定的词，但有时又需要进行阶段性回顾，衡量目前的探究进展。

在科学领域的探究中这一点显得尤为重要，随时调整探究视角、了解不同层级的结合方式，有助于理解复杂系统的运作。例如，了解细胞层级的运作过程，将细胞活动与整个身体的活动联系起来。有时候，退后一步还可能帮助你识别某种模式，这在历史、艺术和科学中可能很重要。

同义动词	相关词汇		智力美德
检查	聚焦	大局	集中
仔细观察	细节	概述	综合
复审	特定的	整体的	
统观	微小的	系统	
	方面	模式	
	视角	完形	

03

ACTING PLAN

S, M, H, D, W, M, Y, D, C, M

First solve the puzzle of what this series means—units of time from second to millennium. Then consider something broad and interesting—the life of an animal, or a nation. Think about what might be important in understanding it at each of these timescales—from respiration to natural selection, or from the signing of a treaty to the preservation of a language.

Yo-Yo Facilitation

Move from a particular question, such as, "Was it right for the UK to go to war over Germany?" to a more general question, such as, "When is it right to go to war?" and back again, using the particular arguments to inform general principles and vice versa. (Also see the Hokey-Kokey Method from the Philosophy Foundation.)

Gradual Picture Reveal

Start with a small detail from a picture to get focused attention and invite predictions as you gradually reveal the whole thing.

03

步法应用

时间尺度

首先解开标题中这串字母的含义——它们代表着从"秒"到"千年"这一系列时间尺度的单位。然后，我们可以思考一些更广泛且有趣的东西，比如，某种动物的生命或某个民族存在的时间。影响动物生命最主要的因素是什么？影响民族存在时间最重要的原因是什么？若从不同的时间尺度上去考量，答案是否不尽相同？对动物的生命来说，可能是短暂的一次呼吸或优胜劣汰的自然选择；对某个民族的兴亡来说，可能是某一次条约的签署或者语言的延续。

悠悠球引导法

从某个特定的问题开始，比如："英国对德国开战是正确的吗？"然后再提出一个更具有普适性的问题，比如："什么时候发动战争是正确的？"最后回过头来，用对特定情况的论证来验证一般性原则，或是用一般性原则来说明特定问题。（参见哲学基金会的 Hokey-Kokey 方法。[①]）

[①] 由彼得·沃利提出的哲学化探究策略。从某个具体问题（故事中或生活中出现的）开始讨论，基于概念提出抽象问题，引导共同探究者对具体问题相关的概念和普适情景进行进一步的思考。最后再回到具体问题之中，重新思考人物或事件情境。——译者注

Half-Time Oranges

At some point through a project or piece of work (it doesn't have to be mid-way), ask learners to pause and consider which Thinking Moves they have made so far and which would be most helpful to do next. You could go a step further by asking which have been easiest or hardest, and activate some members of the class to act as "Move Mentors" for others.

Applications

Applications
Yo-yo facilitation

Geography
Is salmon farming sustainable? What makes something sustainable?

English
Is *Jekyll and Hyde* a typical Gothic novel? What makes a Gothic novel?

Biology
Are bacteria parasites? What makes something a parasite?

Wise Sayings

That's been one of my mantras—focus and simplicity.

——Steve Jobs

See what happens if you step back instead of bounding forward.

——Nora Roberts

渐进式画面显示

从图片局部的某个细节开始，请参与者进行观察和猜测，逐步地渐进地展示，直到画面最终完全呈现，请参与者一步步进行猜测。

橙子的中场休息

在完成项目或工作的某个时间节点（不一定是进程过半时），要求学生暂时停下来，对思考步法的使用进行思考。目前为止，使用了哪些思考步法？哪些思考步法可能对下一步的工作最有帮助？更进一步的提问还可以是：哪些思考步法是最容易的？哪些思考步法是最难的？鼓励一部分学生主导围绕思考步法进行的反思工作，尝试着担任其他人的"思考步法导师"。

学科中的应用

应用

悠悠球引导法

地理

鲑鱼养殖可持续吗？保证可持续性的主要因素是什么？

英语

《化身博士》是典型的哥特式小说吗？构成哥特式小说的关键是什么？

生物

细菌是寄生虫吗？寄生虫的特征是什么？

名人语录

保持专注和简单一直是我的秘诀。

——史蒂夫·乔布斯

试试看不再向前跳跃，如果退后一步，会发生什么？

——诺拉·罗伯茨

思考径

以下是一些在完成学习任务时可以使用的思考步法序列。它们之所以被称为"思考径"，是因为，随着不断地练习和使用，这些思考步法序列会在脑海中根深蒂固——就像一条因不断使用而形成的路径或沟槽。

思考径的想法，与罗恩·里奇哈特所创的便于使用的"思维习惯"类似，但仍存在明显的不同。

其中一个主要的差异是，里奇哈特的思维习惯不是序列式的：它们是单个的动作，且通常由教师进行提示。而思考径则是序列式的，同时，是希望能够由学生自行来使用的。因此，思考径不仅是认知策略，还是一种自主调配的元认知策略。

另一个差异是，由于思考径是由 26 种基本思维过程组合而成的，形成的组合种类多样且富有针对性和目的性。

当然了，这并不是说所有 650 种由 2 个步法组成或所有由 3 个步法组成的 15 600 种思考径都是"有效"的。下面以思维习惯中最广为人知的组合：我看见—我想到—我好奇为例，展示思考步法如何能将思维过程表达得更丰富、更准确。

首先，"看 / 听"比"看见"涵盖了更广义的思维过程。

其次，在上下文提供的语境中，"想到"是宽泛的。在具体的语境中，为保证对思维过程进行描述的精度，它可以用以下步法中的某一种替代：关联、区分（寻找差异）、解释、标题、推测、关键词、保持、描绘（你自己）、回应、聚焦细节。

最后一步，虽然用"好奇"鼓励学生思考是件好事，但如果进入"提问"——让他们更准确、更公开地将自己的好奇组织并表达出来，可能是更有益的。

因此，思考径是一种通用的策略，一些思考径非常适合在日常或学术活动中使用，同时，使用者还可以结合个人情况，组成新的思考步法序列，创造属于自己的思考径，以应对更复杂的情况，比如在进行关键决策或解决冲突时。

这样一来，思考径就为达成更审慎的思考提供了个性化配置的工具箱。

以下是一些在学术场景中可能使用到的思考径。如果您在自己的教学实践中为某个学科或主题开创了新的思考径，非常期待您能与我们分享。我们会在自己的网站上对其进行更新发布。

项目计划阶段

- 度量（估计待办的任务量）
- 权衡（选择）（判断达成目标可以使用的最佳方法）
- 排序（进行任务优先级排序并安排调配）

项目进行阶段

- 试验（验证）(根据目标和时间节点，追踪和检查项目进度）
- 提前思考（预测项目是否能够按时完成）
- 变通（根据目标，及时对实施进行调整）

项目结束后

- 回顾（复盘所取得的进展）
- 提问（询问若重来一次，会进行调整的部分）
- 标题（总结经验和教训）

***

在学校旅行之前

- 关联（将旅行目的地与先前的学习经验进行关联）
- 描绘（想象你在那里可能会发生的事）
- 提问（在旅行中可能的发现或疑惑）

在学校旅行中

- 听/看（留意并能够说出旅行目的地的特点）

- 变焦（有针对性地关注目的地的某些特点或特定事物）

- 推测（推断或猜测特点或事物背后的意涵）

在学校旅行结束后

- 回顾（回顾旅行中的亮点）

- 回应（回答旅行中产生的问题）

- 标题（总结所学内容）

***

规划论文写作

- 区分（罗列论文的内容板块）

- 归类（按主题 / 视角对想法进行分类）

- 排序（将要点按照流畅的逻辑顺序进行排列）

撰写论文中的某个段落

- 构想（提出主要想法或观点）

- 解释（澄清该想法或观点与论文所要解决问题之间的关系）

- 举例（举例说明你的观点）

撰写论文结论

- 权衡（选择）（决定你的最终论点）
- 保持（坚持并说明观点）
- 证明（为自己的选择给出理由）

***

复习

- 回顾（回忆你学过的东西）
- 区分（分列关键术语）
- 解释（根据自己的理解给关键术语下定义）

文本理解

- 关键词（将问题中的关键词标识出来）
- 推测（阅读文本，从文中得出结论）
- 回应（回答问题）

辩论

- 构想（为你方所支持的观点提出理由）

- 证明（为这个理由辩护）

- 回应（回答反对方可能提出的想法）

THINKING GROOVES

These are sequences of Thinking Moves that that can be used for learning tasks. They are so-called because, when practised regularly, they become *ingrained*—like a well-trodden pathway or groove that is formed by continual use.

This idea is similar to Ron Ritchhart's handy Thinking Routines[1] but with significant differences.

A main one is that many of Ritchhart's routines are not sequences: they are single moves, routinely prompted by the teacher. Thinking Grooves are always sequences; but, also, they are intended to be appropriated by students, i.e. used independently. They thus become not just cognitive strategies, but metacognitive strategies—deployed autonomously.

Another point of difference is that, because Thinking Grooves are different combinations of the 26 fundamental processes of thinking, there can be a great variety of them, whilst being very precise and purposive.

Of course, not all of the 650 two-Move combinations, or the 15 600 three-Move combinations, "work", but here is a demonstration of how the Moves can provide both more variety and more precision to one of the

---

1   Ritchhart, R., Church, M. and Morrison, K. (2011). *Making Thinking Visible: How To Promote Engagement, Understanding, And Independence For All Learners*. San Francisco, Ca: Jossey-Bass.

most popular Thinking Routines: SEE-THINK-WONDER.

For a start, LOOK/LISTEN is both more precise than SEE but with twice the potential.

Then THINK is about as imprecise as one could be in the context. It could more meaningfully be replaced by any of: CONNECT, DIVIDE (look for differences), EXPLAIN, HEADLINE, INFER, KEYWORD, MAINTAIN, PICTURE (yourself), RESPOND, ZOOM IN.

And, finally, whilst it is good to encourage students to wonder, it is better still to ask them to QUESTION, both more precisely and more publicly.

Our concept of the Thinking Grooves, then, is not only that there might be ones that are fit for regular use in everyday or academic activities, but also that individuals can themselves invent and implement combinations for more complex situations, such as critical decision-making, or conflict resolution.

Thinking Moves thus provides a sort of do-it-yourself toolkit for more thoughtful activity.

Here are a few suggested Thinking Grooves for academic use. If any teacher invents and uses a new groove for a particular subject or topic, we would be pleased to hear of it and to post it, with due credit, on our website!

Planning a project

- SIZE (measure up the task in hand)
- WEIGH UP (judge the best means to ends)
- ORDER (prioritize and delegate tasks)

During a project

- TEST (check progress against objectives, and time limit)
- AHEAD (predict if the project will be completed by the deadline)
- VARY (make any changes needed to achieve this goal)

After a project

- BACK (reflect on what was achieved)
- QUESTION (ask if you would do anything differently)
- HEADLINE (summarize what you have learnt for next time)

***

Before a school trip

- CONNECT (link the destination with prior learning)
- PICTURE (visualize yourself being there)
- QUESTION (wonder what might be found and found out)

On a trip

- LISTEN (notice and name special features of the destination)
- ZOOM (think about some features or objects in particular)
- INFER (deduce or guess what lies behind the features or objects)

After a trip

- BACK (recall highlights)
- RESPOND (answer questions about what you learnt)
- HEADLINE (summarize what you learnt)

***

Planning Essays

- DIVIDE (list what could go into the essay)
- GROUP (sort the ideas by theme/perspective)
- ORDER (arrange key points into a list of flowing paragraphs)

Writing an essay paragraph

- FORMULATE (suggest an idea or point)
- EXPLAIN (clarify how this relates to the question)
- eXEMPLIFY (give examples to support your point)

Writing an essay conclusion

- WEIGH UP (decide upon your side of the argument)
- MAINTAIN (assert your point of view)
- JUSTIFY (give reasons for your decision)

***

Revision

- BACK (recall what you have learnt)
- DIVIDE (separate key terms)
- EXPLAIN (give their meanings or definitions)

Text comprehension

- KEYWORD (highlight the key words in the question)
- INFER (draw a conclusion from the text)
- RESPOND (answer the question)

Debating

- FORMULATE (suggest an idea for your side)
- JUSTIFY (argue for this idea)
- RESPOND (answer the opposition's ideas)

在学校和生活中
使用思考步法

26 个思考步法中，有 18 个非常适用于讨论，尤其适用于哲学化的探究活动。

熟悉思考步法的教师或引导者能更清楚地把握推进讨论的节奏，何时需要有人**回应**问题，何时需要**证明**某个主张。他们会让学生在最合适的时机及时变换讨论的视角，从某些细节中抽离，回到问题本身，通过**关注整体**来进行思考；或是深入考量，**聚焦**于某个具体的词语或探究路径来加深对问题的认知。

我们建议教师将这类思考步法清晰明确地教授给学生，以帮助他们更好地进行讨论或探究。

当然，一下子掌握 18 个思考步法可能比较困难，我们建议可以将它们分为两组，每组 9 个，从第一组开始：

回顾、关联、证明、关键词、听 / 看、否定、提问、回应、权衡（选择）

我们大致可以按照以下的叙述顺序来记忆它们：

- 听 / 看（对刺激物，或其他人提及的内容给予关注）
- 回顾（回想你之前听到或看到的）
- 关联（与你之前的经验或课程进行关联）
- 提问（对某个字词或想法的意义、真实性和价值，提出质疑）
- 回应（给出回答或评论）
- 否定（不同意某个观点或想法）

- 证明（对你的立场进行论证）

- 关键词（突出重要的想法）

- 权衡（选择）（决定自己的最终立场）

第二组的 9 个是：

区分、解释、构想、标题、保持、试验（验证）、举例、让步、变焦（聚焦细节 / 关注整体）

这组步法大致的叙述顺序是：

- 构想（提出一种新的思路）

- 试验（验证）（对某人所说的内容表示怀疑）

- 保持（坚持某个信念或立场）

- 解释（说明你的思考是什么样的）

- 标题（对他人说到内容进行总结）

- 举例（给出一个例子来阐明或支持自己的立场）

- 区分（区分或分辨不同的观点和立场）

- 变焦：（关注整体）（如调查其他人说过的）

- 让步（接受其他人的某个观点）

我们建议引导者将上述的 18 个步法及括号中的解释制成 A5 大小的表格形式，放置在讨论 / 探究环境之中，这样参与者可以随时参阅，能觉察到自己或同伴在思考过程中正在使用的思考步法。（第二组也可被用于"金鱼缸"式的课堂观察记录中，这对元认知发展来说将是非常好的练习。）

另外还有更多步法未被列出，它们大部分也能在探究中使用，形成更有深度、更具批判性的想法。但通常来说，它们在日常教学或生活场景中的应用可能更为广泛，以下是第三组步法：

提前思考，归类，推测，排序，描绘，度量，使用（工具），变通，变焦（聚焦细节）

想要为这几个步法排出通用的叙述性顺序似乎不那么容易，我们试着以教室为应用场景给出一个建议：

- 提前思考（预测行动的后果）

- 归类（进行分类或分级）

- 推测（推出结论或含义）

- 排序（按优先级排列事物）

- 描绘（可视化，甚至设身处地将自己放置到某个场景之中）

- 度量（估计一个主张或问题的范围）

- 使用（工具）（在课堂上尝试一个提议或进行一个"思想实验"）

- 变通（改变视角或关注点）

- 变焦（聚焦细节）（关注某个关键词或立场）

在本节的最后我们需要说明的是，虽然这本书主要是为教师和课堂使用而写的，但这些思考步法并不仅仅是一种学术工具，它们几乎涵盖了人们所有的思维过程。因此，这些思考步法同样适用于日常用途，如组织活动或经营企业。具体地说，委员会主席主持讨论会和教师引导课堂探究所需要的技能可能是非常相似的。

USING THINKING MOVES IN SCHOOL AND LIFE

Eighteen of the Moves are particularly well suited for discussions, especially philosophical inquiries.

A teacher or facilitator who knows the Moves well will see more clearly when there is a need for someone to RESPOND to a question or to JUSTIFY a claim. She or he might pick the best moment to ask everyone to ZOOM OUT and think BACK to the question, or perhaps to ZOOM in on a particular word or avenue of inquiry.

We recommend that such Moves be explicitly taught to students for the purposes of discussion/inquiry.

Eighteen is rather a lot to digest at once, of course, so we suggest splitting them into two groups of nine, starting with:

BACK, CONNECT, JUSTIFY, KEYWORD, LISTEN/LOOK, NEGATE, QUESTION, RESPOND, WEIGH UP

It might be easier to think of them in a rough narrative order:

- LISTEN (pay attention to a stimulus—or to what people say)
- BACK (recall what you have heard or seen)
- CONNECT (to other experiences or lessons you have had)
- QUESTION (the meaning, truth and value of words or ideas)

- RESPOND (with answers or comments)
- NEGATE (disagree with a belief or position)
- JUSTIFY (argue for your position)
- KEYWORD (highlight important ideas)
- WEIGH UP (decide on your latest position)

The next nine, then, would be:

DIVIDE, EXPLAIN, FORMULATE, HEADLINE, MAINTAIN, TEST, eXEMPLIFY, YIELD, ZOOM (OUT)

Here is a rough narrative order for these:

- FORMULATE (suggest a new way of thinking)
- TEST (put what someone has said in doubt)
- MAINTAIN (commit to a belief or position)
- EXPLAIN (say how you think things are)
- HEADLINE (summarize what you think others are saying)
- eXEMPLIFY (give an example to clarify or support your position)
- DIVIDE (separate, or distinguish between, ideas and positions)
- ZOOM (out—i.e. survey what people have said)
- YIELD (accept others' point of view)

Each of these 18 Moves, with the bracketed explanations, is on an A5 sheet that can be downloaded from the support resources. We would recommend that facilitators distribute these so that discussants/inquirers can look for opportunities to "make the Move", or notice when the Move is made by others. (The latter can also be done by observers in an outer circle or fishbowl arrangement, which would be good practice for metacognition.)

There are more Moves not listed above. Most of them can also be used in inquiry, for developing deeper, more critical thinking, but some of them may be regarded as useful more widely, in routine teaching or in everyday life. So, here is the third set:

AHEAD, GROUP, INFER, ORDER, PICTURE, SIZE, USE, VARY, ZOOM (IN)

These do not so easily form a narrative order, but here are suggestions as to how they could also be useful in the classroom:

- AHEAD (predicting consequences of actions)
- GROUP (sorting or classifying)
- INFER (deducing conclusions or implications)
- ORDER (putting things in order of priority)
- PICTURE (visualizing, or even putting yourself in someone's shoes)
- SIZE (estimating the scale of a claim or of a problem)
- USE (trying out a proposal in class or perhaps a "thought experiment")

- VARY (changing perspective or focus)
- ZOOM (in—focusing on a key word or proposition)

A final thought for this section: whilst this book is written primarily for teachers and use in classrooms, the Moves are not a merely academic device. They include almost all the possible moves that any thinker could make in any circumstance. So, they are equally appropriate for everyday purposes, such as organizing events or running businesses. The skill set needed by a chair of a committee, for example, is very similar to that of a teacher conducting a classroom enquiry.

如何记忆思考步法

更关注思维培养的教育工作者，常常被误解为是反对教授"知识"的，仿佛思考或教学本身是不包含任何知识的！（只要想想为进行基本的逻辑思考我们所需要了解的词汇，比如"推测"或"证明"，就能体会到上述论点的草率。）

我们不仅认同知识教授是进行良好思考和判断的基础，还认可有时候死记硬背也是有价值的。事实上，我们相信记住这些思考步法对日常生活和学术发展都具有很大价值。因为记忆是熟练运用的基础，这既能让自己更熟练地运用它们，又能帮助你更清晰地从他人的活动中识别它们（有点像德博诺的思考帽，或者他的便捷记忆法，利弊兴趣分析法、目标导向法、全面考虑法等）。

我们认为，不管是几岁的头脑都可以快速并彻底地记住这套思考步法。

26个思考步法除了以 A—Z 的首字母顺序排序使其变得易于记忆之外，它们被设置为成对出现的，这样几乎让记忆的工作量减半。某几对甚至具有首字母连续性：

提前思考 / 回顾（在时间上）

关联 / 区分（在空间上）

解释 / 构想 / 试验（验证）（一个想法）

推测 / 证明（一个结论）

保持 / 否定（一个主张）

提问 / 回应（显而易见？）

度量 / 试验（验证）（一个要求）

使用（工具）/ 变通（一个想法）

变焦（聚焦细节 / 关注整体）（算是变焦的两种形式）

其他的几对步法虽然有些间隔，但也很适合配对：

归类 / 举例（和概念有关的事物）

标题 / 关键词（主要观点）

听 / 看 / 描绘（词语 / 世界）

度量 / 排序（项目）

权衡（选择）/ 变通（考量）

讨论这些对子关联的相关形式的不同方式（如反义词、补语、变体）不仅可以帮助学生记住这些动作，还可以帮助学生更好地理解和欣赏它们。

1. 我们建议您向学生提供思考步法清单（即本书中的思考步法 A—Z 及关键同义词列表（包含 26 个思考步法及其关键同

义词），您也可以从网站上下载这一清单。

2. 快速浏览思考步法列表，让学生勾选自己听说过的思考步法以及思考步法同义词。此时不用对每个步法展开解释。

3. 如果大多数学生对某个思考步法不太熟悉（学生可能不太熟悉的步法是：构想、推测、证明、保持、否定、度量、试验（验证）、举例、让步和变焦。请他们查看步法的关键同义词，勾选出其中听说过的词。如果学生对同义词也不熟悉的话，此时教师可以花时间解释一下该思考步法。

4. 从列表中的 A 开始，向学生介绍以上关于步法的配对。如果有需要，可以花些时间讨论它们。

5. 在第一节思考步法课结束时，给每位学生一个空白的A—Z 思考步法表（可从网站下载）。两两结对，他们需要尽可能多地回忆并填写步法（不查看填写完整的思考步法主表）。然后告诉他们在本周结束时将以同样形式进行另一次关于思考步法的书面测验。教师应该尽量让这个过程变得有趣而不成为学生的负担。您可以为表现最好的组准备小奖品，请期待每个小组的成功，为他们准备足够的奖品哦！

在开始学习思考步法的最初几周内，如果能够以每周 2—3 次的频率进行强化巩固，对学生记忆和掌握思考步法将非常有效。要尽快从书面测试转到口头背诵 A—Z，然后尝试倒过来

如何记忆思考步法

从 Z—A 进行背诵。（不要害羞，可以全班一起背诵！）在两次思考步法测试期间，教师尽可能多地在所有课程中示范思考步法，在语言和行为上提及并运用思考步法。这样，教师和学生很快就会对 26 个思考步法非常熟悉。您也可以在每节课中指定几位学生扮演"思考步法监督员"，他们的职责是识别出在每节课中被使用的思考步法及其使用者。

最后的小提示：如果希望您的学生能尽快"掌握"这些思考步法，一个简单的方法是每节课给每个学生分配一个不同的思考步法，他／她可以有针对性地关注该步法在整节课中的使用。理想情况下，他／她会记得是谁／在什么情况下使用了该思考步法。虽然课堂活动时间有限，教师无法在每节课的最后反思所有思考步法的监控情况，但您可以每次选择 1—2 个（尤其是那些不常见或不易识别的步法）来进行反思回顾。您也可以鼓励学生在其他课程或日常生活中识别和使用思考步法，给他们一些小奖励。尽早地组织和鼓励这些关于思考步法的反思／识别活动，学生在未来的实施和应用时就会越来越得心应手。

How to Memorize
the Moves

Educators who advocate more of a focus on the explicit teaching of thinking are often misrepresented as being against the teaching of "knowledge"—as if there is no knowledge, anyway, about either thinking or teaching! (One only has to think of the vocabulary, such as Infer or Justify, that one needs to know in order to perform even basic moves of logic, to appreciate the strawness of that argument.)

We not only accept the need to teach knowledge as a basis for good thinking and judgement, but also accept rote learning can sometimes be a good thing. Indeed, we believe that learning the moves by heart would be of great value for both everyday and academic purposes. This is because having them ready to mind enables one both to employ them more skillfully oneself, but also to elicit them more systematically from others (a bit like de Bono's Thinking Hats[1], or his handy mnemonics, PMI, AGO, CAF, etc.)

We maintain, moreover, that young minds, and even older ones, can memorize the moves very quickly and thoroughly.

Apart from the fact that the alphabet provides an easy series of pegs on which to hang the moves, they also come in a number of pairings,

---

1 De Bono, E. (1985). *Six Thinking Hats*. London: Penguin Life. 中译本见爱德华·德博诺著，马睿译《六顶思考帽——如何简单而高效地思考》，北京：中信出版集团，2016 年。

which almost halve the job of memorizing. Some pairs even have the convenience of being consecutive, as follows:

Ahead/Back (in time)
Connect/Divide (in space)
Explain/Formulate/Test (an idea)
Infer/Justify (a conclusion)
Maintain/Negate (a claim)
Question/Respond (obviously?)
Size/Test (a claim)
Use/Vary (an idea)
Zoom in/Zoom out (counting ZOOM twice)

Others pair quite well at a distance:

Group/eXemplify (things by concepts)
Headline/Keyword (main ideas)
Listen/Look/Picture (words/worlds)
Size/Order (items)
Weigh up/Yield (considerations)

Discussion of the different ways in which these pairs connect (e.g. opposites, complements, varieties) can help students not only remember the moves but also understand and appreciate them better.

1. We recommend you give students a copy at once of the master sheet of moves (i.e. with their key partners) that can be found in this book. You can download the sheet from the website.

2. Have a quick run through of the list, asking students to tick off every move they have heard of before, along with the synonyms they've heard of. Don't worry about explaining each move at this point.

3. If a move is not familiar to most students (the most likely are: FORMULATE, INFER, JUSTIFY, MAINTAIN, NEGATE, SIZE, TEST, eXEMPLIFY, YIELD And ZOOM), ask them to look at the synonyms and tick off one or both if they've heard of them. Take time to explain a move if the synonyms, too, are unfamiliar to some.

4. Go back to the start of the list and point out the pairings above. Spend time discussing them if it's helpful.

5. End this first session by giving each student a blank A—Z (download from the website). In pairs, they should fill in as many of the moves as they can (without looking at their master sheets). Then tell them they will be given another written quiz in pairs at the end of the week. Make the process fun rather than a chore. If suitable, you could offer prizes for the most successful pairs. Just be prepared with enough prizes for everyone to succeed!

Learning the moves is most effective if it is reinforced two or three time in the first few weeks—so move as quickly as you can from the written test to oral reciting of the A—Z, then perhaps try the Z—A.(Don't be shy of whole class reciting!) Refer to the moves as often as you can in lessons in between testing. They are intended for regular mention and mobilization. Before long, you and your students will be so familiar with them that you might appoint a couple of "Moves Monitors" each lesson, whose job it is to identify the use and user of several moves per lesson.

A final tip: it would be desirable for your students to "own" the moves as quickly as they can. A simple way of doing this would be to allocate a different move each lesson to each student. S/He could then monitor the use of that particular move. Ideally, s/he would note down the name of the user as well as the context of its use. You won't have time to have all the moves made public at the end, but you could check off one or two each time, especially those less common or less easy to identify. You could give bonuses to students who report using "their" move in other lessons or in their everyday lives. Time invested in such reporting early on will pay dividends in students' increasing competence in implementing the moves in future.

思考步法与其他
"关于思考的方案"
的联系

30 年来，罗杰·萨特克利夫一直关注并投身于儿童思考技能的培养活动，对儿童哲学（P4C）投入尤深。儿童哲学（P4C）的创始人马修·李普曼教授，仍然是该领域最伟大的贡献者之一，在其著作《教育中的思维》和自传《教学思维的一生》中深刻阐述了他对儿童思考的重视。

然而，随着实践的深入，罗杰·萨特克利夫越来越关注思考技能的模糊性和争议性。他也发现，自己将 P4C 作为在学校中发展思维的一种方式进行推广和实施，却没有明确意识到——实际上他所试图推广的是一个清晰、简单的关于思考不同类型的思维清单。

罗杰·萨特克利夫将马修·李普曼教授的"批判式、创新式和关爱式思维"的思维框架拓展细化为广泛使用的 4C 框架（批判式、创新式、关爱式和合作式思维），通过更详细的阐述，广泛运用到了儿童哲学活动实施过程中（合作式思维是第四条"腿"）；当然，像许多 P4C 实践者一样，他借鉴了李普曼教授的两位主要同事安·玛格丽特·夏普和劳伦斯·斯普利特的《更好的思考教学》。

关于思考的研究很多，也有很多关于思考技能的清单。2005 年由大卫·莫斯利领导的团队所出版的《思考框架》回顾了不少于 42 种的思维模式。但这个框架不太符合教师的需要，

思考步法与其他"关于思考的方案"的联系

一线教师需要的是一个清晰、容易向学生解释，也容易让学生记住的思考技能清单。

2010年，罗杰·萨特克利夫参加了由阿特·科斯塔主持的关于思维习惯的工作坊，他产生了构建一个新的、容易记忆且涵盖不同思考行为清单的设想。

可以说，最著名的方案——布鲁姆分类法事实上满足了这些需求，它只有6个主要类别，相当容易掌握。但是，除了其明显的等级结构争议，以及对探究、论证和想象等重要范畴的忽略之外，还应指出的是，它更像是一个仅面向教师，而非能够同时面向学生的方案，洛林·安德森和大卫·克拉斯沃尔在2001年出版的包含19个子类别的修订版《教育目标：认知领域的分类》亦然。

我们所需要的是一种思考自己的思维过程的思考方式，具备日常操作性、有条理的、令人难忘的方式，而不是充满专业术语的理论建构。这就是构想思考步法A—Z时的初衷，它是通过对大量思考实践的常态化反思，从元认知的角度来进行构建的。

从一开始，思考步法就使用了动作性而不是技能型的语言来描述思维过程，作为一种要采取的步骤，而不是像"更具创造性的思考"这样通用的概念或指令，这样它就可以更精确地分析整个思考的过程。

最后，罗杰·萨特克利夫将 26 个思考步法以英文字母表的顺序进行了排列，用一种最自然的助记法来帮助学习思考步法。这不是一个简单或短暂的项目，因为一旦为某个步法匹配了字母，任何以同一字母开头的步法都必须作为其他步法的同义词而存在。完成这项任务就像拼一幅巨大的拼图。

CONNECTIONS WITH OTHER THINKING SCHEMES

Beginning 30 years ago, Roger became more and more involved in the Thinking Skills movement—and especially Philosophy for Children (P4C)[1]. Professor Matthew Lipman, the author of P4C, remains one of the great contributors to the field, not least through *Thinking in Education*[2], and his autobiography, *A Life Teaching Thinking*[3].

Roger became increasingly concerned, though, about the vagueness and controversies surrounding Thinking Skills. He became concerned, also, that he was promoting P4C as a way of developing thinking in schools without a clear sense—indeed, list—of the different sorts of thinking he was trying to promote.

To be fair, he had elaborated Lipman's framework of Critical, Creative and Caring Thinking into the widely used 4C framework (with Collaborative Thinking as the fourth "leg"); and, like many a P4C practitioner, drew on the work[4] of Ann Margaret Sharp and Laurance Splitter, two of Lipman's leading associates, for a provisional list.

---

1 Institute for the Advancement of Philosophy for Children-https://www. montclair.edu/iapc/.

2 Lipman, M. (2003). *Thinking in education*. Cambridge Etc.: Cambridge University Press. 中译本见马修·李普曼著，刘学良、汪功伟译《教育中的思维——培养有智慧的儿童》，上海：华东师范大学出版社，2023 年。

3 Lipman, M. (2008). *A life teaching thinking*. Montclair, N.J.: Published By The Institute for the Advancement of Philosophy for Children.

4 Splitter, L.J., Ann Margaret Sharp and Australian Council For Educational Research (2005). *Teaching for better thinking: the classroom community of inquiry*. Melbourne, Vic: Acer.

CONNECTIONS WITH OTHER THINKING SCHEMES

There were, actually, many lists of thinking skills. "Frameworks for Thinking"[1], published in 2005 by a team led by David Moseley, reviewed no fewer than 42 such schemes. Yet none were what teachers needed them to be: easy to explain to students and easy for students to memorize.

It was when Roger attended a workshop on Habits of Mind, led by Art Costa[2] in 2010 that he conceived of the possibility of constructing a memorable list of distinct acts of thinking.

It could be argued that the most famous scheme, Bloom's taxonomy, fulfilled these needs, having just six main categories that were fairly easy to grasp. But, quite apart from controversies about its apparently hierarchical structure, and its omission, on the face of it, of vital categories such as inquiry, argumentation, and imagination, it has been more of a teachers' than a students' scheme, as has the 19-sub-categories version[3] published by Lorin Anderson and David Krathwohl in 2001.

What was needed was an everyday, methodical and memorable way of

---

1   Moseley, D. and Al, E. (2005). *Frameworks for thinking: a handbook for teaching and learning*. Cambridge, UK; New York: Cambridge University Press.

2   https://www.habitsofmindinstitute.org/about-us/.

3   Anderson, L.W. and Krathwohl, D. (2001). *A Taxonomy for Learning, Teaching, and Assessing: A Revision of Bloom's Taxonomy of Educational Objectives*. Essex: Pearson. 中译本名称不同，见洛林·W. 安德森等编著，蒋小平等译《布鲁姆教育目标分类学——分类学视野下的学与教及其测评》，北京：外语教学与研究出版社，2009 年。

thinking about one's thinking, without resort to technical terms. That is how the A—Z was conceived and it was constructed through regular reflection on thinking in practice—through metacognition, no less!

From the start it used the language of Moves rather than skills, so that it could analyze thinking more precisely as clear steps to be taken, rather than general concepts or injunctions, such as "think creatively".

Finally, Roger mapped the Moves methodically to the letters of the alphabet—employing the most natural mnemonic to assist in learning the Moves. This was no easy or short project, as once a letter was allocated to a particular Move, any other Move beginning with the same letter had to be subsumed as a synonym of a different Move. The task was similar to that of solving a gigantic jigsaw puzzle.

图书在版编目（CIP）数据

思考步法A-Z：AI时代元认知入门 / (英) 罗杰·萨特克利夫，(英) 汤姆·比格尔斯通，(英) 贾森·巴克利著；儿童哲学·中国译. — 上海：上海教育出版社，2025.4. —（世界儿童哲学文库）. — ISBN 978-7-5720-3392-6

Ⅰ. B842.1

中国国家版本馆CIP数据核字第20259XX238号

策划编辑　刘美文　王　璇
责任编辑　王　璇
插　　画　安静Echo
装帧设计　王鸣豪

Thinking Moves A–Z: Metacognition Made Simple
Written by　Roger Sutcliffe　Tom Bigglestone　Jason Buckley

思考步法A–Z——AI时代元认知入门
[英] 罗杰·萨特克利夫　[英] 汤姆·比格尔斯通　[英] 贾森·巴克利　著
儿童哲学·中国　译

出版发行　上海教育出版社有限公司
官　　网　www.seph.com.cn
地　　址　上海市闵行区号景路159弄C座
邮　　编　201101
印　　刷　苏州工业园区美柯乐制版印务有限责任公司
开　　本　889×1194　1/32　印张 10.75　插页 4
字　　数　186 千字
版　　次　2025年4月第1版
印　　次　2025年4月第1次印刷
书　　号　ISBN 978-7-5720-3392-6/B·0091
定　　价　68.00 元

如发现质量问题，读者可向本社调换　电话：021-64373213

扫码解锁更多关于
思考步法及儿童哲学数字资源